**Kermit Sigmon**

CRC Press
Boca Raton Boston New York Washington, D.C. London

The graphics on the cover (a model of the familiar Mobius strip, shown here with five loops (front) and perspective and angle control combined with lighting and interpolated shading highlight the New England coast (back). Data courtesy of NOAA. Both were created with MATLAB 5 and are provided courtesy of The MathWorks Inc., Natick, MA



**Library of Congress Cataloging-in-Publication Data**

Catalog record is available from the Library of Congress.



Direct all inquiries to CRC Press LLC, 2000 N.W. Corporate Blvd., Boca Raton, Florida 33431.


International Standard Book Number 0-8493-1305-6
Printed in the United States of America 3 4 5 6 7 8 9 0
Printed on acid-free paper

# Preface to the Fifth Edition of the MATLAB Primer

The fifth edition introduces a dramatic change in the *MATLAB Primer.* Kermit Sigmon, author of the *MATLAB Primer* and a long-time supporter and friend of The MathWorks, passed away in January 1997. We are grateful to him for the fine contribution he has made to thousands of MATLAB users during the past six years by making this resource available.

Now, we at The MathWorks want his work to continue. We have updated the MATLAB Primer to reflect the new features of MATLAB 5. It is our honor and pleasure to deliver this book and to contribute the proceeds to Professor Sigmon's family in his memory.

Those of us at The MathWorks who collaborated on this effort: Tom Bryan, Peter Costas, Scott Gray, Roy Lurie, Naomi Bulock, Kathy Ford, Jim Tung, Liz Callanan, Mary Ann Branch and I are pleased to extend the work that Professor Sigmon started. Our thanks goes to Bob Stern, Executive Editor in Mathematics and Engineering at CRC Press, for carrying the torch to make this new version a reality.

Cleve Moler

*Chief Scientist and Chairman of the Board*
The MathWorks, Inc.

# Introduction

MATLAB,® developed by The MathWorks, Inc., integrates computation, visualization, and programming in a flexible, open environment. It offers engineers and scientists an intuitive language for expressing problems and their solutions mathematically and graphically. Complex numerical problems can be solved in a fraction of the time required with a programming language such as FORTRAN or C.

The purpose of this Primer is to help you begin to use MATLAB. It is not intended to be a substitute for the *User's Guide* and online *Reference Guide* for MATLAB. The Primer can best be used hands-on. You are encouraged to work at the computer as you read the Primer and freely experiment with examples. This Primer, along with the online help facility, usually suffices for students in a class requiring use of MATLAB.

You should liberally use the online help facility for more detailed information. There are several ways to access online help about MATLAB:

| | |
|---|---|
| `help` | lists topics for which online help is available |
| `help` *topic* | lists those specific functions under this topic for which help is available |
| `help` *functionname* | displays information about a specific function |
| `helpwin` | help window interface |

| | |
|---|---|
| `helpdesk` | uses an Internet Browser to search help information and access online documentation |
| `lookfor` *keyword* | searches the first line of help text in a function based on a keyword |

For example, the command `help eig` will give information about the eigenvalue function `eig`. See the list of functions in the last section of this Primer for a brief summary of help for a function. You can also preview some of the features of MATLAB by first entering the command `demo` and then selecting from the options offered.

The scope and power of MATLAB go far beyond this Primer. Eventually you will want to consult The MathWorks documentation *Getting Started with MATLAB* and *Using MATLAB,* which are available online. Copies of the documentation are also often available for review at locations such as consulting desks, terminal rooms, computing labs, and the reserve desk of the library. Consult your instructor or your local computing center to learn where this documentation is located at your institution.

MATLAB is available for a number of environments: Sun/HP/DEC/IBM/SGI Unix workstations, 486/Pentium computers under MS Windows 95 and NT 4.0, and Apple Macintosh. Although MATLAB was designed to be multi-platform there are a few differences. For example, a built-in M-file editor is available on the Macintosh and PC which provides basic text editing operations as well as access to M-file debugging tools. The information in these notes applies generally to all of these environments.

For more information on MATLAB contact:

The MathWorks, Inc.
24 Prime Park Way
Natick, MA, 01760-1500
Tel: 508–647–7000
Fax: 508–647–7001
Email: info@mathworks.com
WWW: http://www.mathworks.com

A Student Edition of MATLAB is available from Prentice Hall.

# Table of Contents

# 1. Accessing MATLAB

On most systems, after logging in you can enter MATLAB with the system command `matlab` and exit MATLAB with the MATLAB command `quit` or `exit`. However, your local installation may permit MATLAB to be accessed from a menu or by clicking an icon.

On systems permitting multiple processes, such as a Unix system or MS Windows, you will find it convenient, for reasons discussed in section 12.1, to keep both MATLAB and your local editor active. If you are working on a platform that runs processes in multiple windows, you will want to keep MATLAB active in one window and your local editor active in another.

Consult your instructor or your local computer center for details of the local installation.

# 2. Entering Matrices

There are six fundamental data types (or classes) in MATLAB, each one a multidimensional array. The six classes are double, char, sparse, uint8, cell, and struct. The class that we will concern ourselves with most in this text is a rectangular numerical array with possibly complex entries. An array of this type is called a matrix. A matrix with only one row or one column is called a vector. A 1-by-1 matrix is called a scalar.

Arrays can be introduced into MATLAB in several different ways:

- Entered by an explicit list of elements
- Generated by built-in statements and functions
- Created in a diskfile with your local editor
- Loaded from external data files or applications (see *Using MATLAB*, or online `help`)

For example, either of the statements

```
A = [1 2 3; 4 5 6; 7 8 9]
```

and

```
A = [
1 2 3
4 5 6
7 8 9]
```

creates the obvious 3-by-3 matrix and assigns it to a variable *A*. Try it. The elements within a row of a matrix may be separated by commas as well as a blank. When listing a number in exponential form (e.g., 2.34e–9), blank spaces must be avoided.

## 2.1 Complex Numbers

MATLAB allows complex numbers in most of its operations and functions. Two convenient ways to enter complex matrices are:

```
A  = [1 2;3 4] + i*[5 6;7 8]
A  = [1+5i 2+6i;3+7i 4+8i]
```

When listing complex numbers (e.g., $2 + 6i$) in a matrix, blank spaces must be avoided. Either *i* or *j* may be used as the imaginary unit. If, however, you use $i$ and $j$ as variables and overwrite their values, you may generate a new imaginary unit with, say, `ii = sqrt(-1)`.

## 2.2 Large Matrices

Listing entries of a large matrix is best done in an ASCII file with your local editor, where errors can be easily corrected. The file should consist of a rectangular array of just the numeric matrix entries. If this file is named, say, `data.ext` (where `.ext` is any, possibly absent, extension except `.mat`), the MATLAB command `load data` reads this file to the variable *data* in your MATLAB workspace. Large matrices may also be entered with a script file (see section 10.1).

## 2.3 Multidimensional Arrays

Arrays of arbitrary dimension can be formed. For example, `A = zeros(3, 5, 4, 2)` creates a four-dimensional array of size 3-by-5-by-4-by-2. Multidimensional arrays may also be built up using `cat` (short for concatenation).

## 2.4 Using `rand`, `magic`, and `hilb`

The built-in functions `rand`, `magic`, and `hilb`, for example, provide an easy way to create matrices with which to experiment. The command `rand(n)` creates an $n \times n$ matrix with randomly generated entries distributed uniformly between 0 and 1 while `rand(m,n)` creates an $m \times n$ one ($m$ and $n$ denote, of course, positive integers). `magic(n)` creates an $n \times n$ matrix that is a magic square (rows, columns, and diagonals have common sum); `hilb(n)` creates the $n$ x $n$ Hilbert matrix, the king of ill-conditioned matrices. Matrices can also be generated with a for loop (see section 6.1).

## 2.5 Referencing Individual Entries

Individual matrix and vector entries can be referenced with indices inside parentheses in the usual manner. For example, $A(2, 3)$ denotes the entry in the second row, third column of matrix $A$ and $x(3)$ denotes the third coordinate of vector $x$. Higher dimensional arrays are similarly indexed. Try it. A matrix or a vector accepts only positive integers as indices.

## 2.6 Other Data Types (Classes)

Character arrays (strings) are stored efficiently in the char data type. Strings are formed by enclosing text in single quotes:
`s = 'I love MATLAB'`.

Sparse matrices are those with mostly zero entries. MATLAB has efficient methods of operating on sparse matrices. See `help sparse`, and `help full`.

Images are stored efficiently using the `uint8` (8-bit unsigned integer) data type.

Cell arrays are collections of other arrays and are formed using curly braces. For example,
`c = {1:5, 'I love MATLAB', magic(3)}`
creates a cell array.

You may create additional data objects and classes using overloading (see `help class`).

# 3. Matrix Operations

The following matrix operations are available in MATLAB:

| | |
|---|---|
| + | addition |
| – | subtraction |
| * | multiplication |
| ^ | power |
| ' | transpose (real) or conjugate transpose (complex) |
| .' | transpose (real or complex) |
| \ | left division |
| / | right division |

These matrix operations apply, of course, to scalars (1-by-1 matrices) as well. If the sizes of the matrices are incompatible for the matrix operation, an error message will result, except in the case of scalar-matrix operations (for addition, subtraction, division and multiplication in which case each entry of the matrix is operated on by the scalar.

Entry-wise operations are discussed in section 3.2. Relational operations are discussed in section 6.

## 3.1 Matrix Division

The "matrix division" operations deserve special comment. If $A$ is an invertible square matrix and $b$ is a compatible column vector, or respectively a compatible row vector, then

$x = A\backslash b$ is the solution of $A * x = b$ and

$x = b/A$ is the solution of $x * A = b$

In left division, if $A$ is square, then it is factored using Gaussian elimination, and these factors are used to solve $A * x = b$. If $A$ is not square, it is factored using Householder orthogonalization with column pivoting and the factors are used to solve the under- or over-determined system in the least squares sense. Right division is defined in terms of left division by $b/A = (A' \backslash b')'$.

### 3.2 Entry-wise Operations

The matrix operations of addition and subtraction already operate entry-wise, but the other matrix operations just given do not; they are matrix operations. It is important to observe that these other operations, *, ^, \, and /, can be made to operate entry-wise by preceding them by a period. For example, either `[1,2,3,4].*[1,2,3,4]` or `[1,2,3,4].^2` will yield `[1,4,9,16]`. Try it. This is particularly useful when using MATLAB graphics.

## 4. Statements, Expressions, Variables

MATLAB is an *interpreted* language; the expressions you type are interpreted and evaluated. MATLAB statements are usually of the form

*variable* = *expression*, or simply
*expression*

Expressions are usually composed from operators, functions, and variable names. Evaluation of the expression produces a matrix, which is then displayed on the screen or assigned to a variable for future use. If the variable name and = sign are omitted, a variable `ans` (for answer) is automatically created to which the result is assigned.

A statement is normally terminated with the carriage return. However, a statement can be continued to the next line with three or more periods ( ... ) followed by a carriage return. On the other hand, several statements can be placed on a single line if they are separated by commas or semicolons.

## 4.1 Suppressing Display of Results

If the last character of a statement is a semicolon, display of the result is suppressed, but the assignment is carried out. This is essential in suppressing unwanted display of intermediate results.

## 4.2 Case-Sensitivity

MATLAB is case-sensitive in the names of commands, functions, and variables. For example, `solveUT` is not the same as `solveut`.

## 4.3 Listing and Clearing Variables and M-files

The command `who` (or `whos`) lists the variables currently in the workspace and `inmem` lists the compiled M-files currently in memory. A variable or function can be cleared from the workspace with the command `clear` *variablename* (or *functionname*). The command `clear` alone clears all non-permanent variables, whereas the command `clear functions` removes all compiled M-files from memory.

## 4.4 Runaway Process, Machine Epsilon

The permanent variable `eps` (epsilon) gives the machine unit roundoff, about $10^{-16}$ on most machines. It is useful in specifying tolerances for convergence of iterative processes.

A runaway display or computation can be stopped on most machines without leaving MATLAB with **Ctrl–C**.

## 4.5 Saving a Session

When you log out or exit MATLAB all variables are lost. However, invoking the command `save` before exiting causes all variables to be written to a machine-readable diskfile named `matlab.mat`. When you later reenter MATLAB, the command `load` will restore the workspace to its former state. Commands `save` and `load` take file names and variable names as optional arguments (see `help save` and `help load`).

## 4.6 Hardcopy

Hardcopy is most easily obtained with the `diary` command. The command

`diary` *filename*

causes what appears subsequently on the screen (except graphics) to be written to the named diskfile (if the filename is omitted, it is written to a default file named `diary`) until one gives the command `diary off`; the command `diary on` causes writing to the file to resume. When finished, you can edit the file as desired and print it out on the local system. On most systems, the

!–feature (see section 12.1) lets you edit and print the file without leaving MATLAB.

For hardcopy of graphics, see section 15.10.

## 5. Matrix Building Functions

Convenient matrix building functions are:

| | |
|---|---|
| `eye` | identity matrix |
| `zeros` | matrix of zeros |
| `ones` | matrix of ones |
| `diag` | create or extract diagonals |
| `triu` | upper triangular part of a matrix |
| `tril` | lower triangular part of a matrix |
| `rand` | randomly generated matrix |
| `hilb` | Hilbert matrix |
| `magic` | magic square |
| `toeplitz` | Toeplitz matrix |

For example, `zeros(m,n)` produces an $m$-by-$n$ matrix of zeros and `zeros(n)` produces an $n$-by-$n$ one. If $A$ is a matrix, then `zeros(size(A))` produces a matrix of zeros having the same size as $A$.

If $x$ is a vector, `diag(x)` is the diagonal matrix with $x$ down the diagonal; if $A$ is a square matrix, then `diag(A)` is a vector consisting of the diagonal of $A$. What is `diag(diag(A))`? Try it.

Matrices can be built from blocks. For example, if $A$ is a 3-by-3 matrix, then

```
B= [A,zeros(3,2);zeros(2,3),eye(2)]
```

will build a certain 5-by-5 matrix. Try it.

## 6. Control Flow Statements

In their basic forms, these MATLAB flow control statements operate like those in most computer languages.

### 6.1 Variable Controlled Loops (`for`)

For example, for a given $n$, the statement

```
x =[]; for i = 1:n, x=[x,i^2], end
```

or

```
x = [];
for i =1:n
   x = [x, i^2]
end
```

produces a certain $n$-vector and the statement

```
x = []; for i = n:-1:1, x=[x,i^2], end
```

produces the same vector in reverse order. Try them. Note that a matrix may be empty (such as `x = []`).

The statements

```
for i = 1:m
   for j = 1:n
      H(i, j) = 1/(i+j-1);
   end
end
H
```

produce and print to the screen the $m$-by-$n$ Hilbert matrix (try `type hilb` to see a more efficient way to produce the same matrix). The semicolon on the inner statement is essential to suppress printing of unwanted intermediate results. The last `H` displays the final result.

The `for` statement permits *any* matrix to be used instead of 1:n. The variable *c* just consecutively assumes the value of each column of the matrix. For example,

```
s = 0;
for c = A
   s = s + sum(c);
end
```

computes the sum of all entries of the matrix *A* by adding its column sums (of course, `sum(sum(A))` does it more efficiently; see section 7.2). In fact, since `1:n = [1,2,3,...,n]`, this column-by-column assignment is what occurs with `for i = 1:n`.

## 6.2 Relation Controlled Loops (`while`)

The general form of a `while` loop is

```
while relation
   statements
end
```

The statements will be repeatedly executed as long as the relation remains true. For example, for a given number *a*, the following computes and displays the smallest nonnegative integer *n* such that $2^n > a$:

```
n = 0;
while 2^n < a
   n = n + 1;
end
n
```

(See `help log2` for a more efficient way to achieve the same result.)

## 6.3 Branching (`if`)

The general form of a simple `if` statement is

```
if relation
   statements
end
```

The statements will be executed only if the relation is true. Multiple branching is also possible, as is illustrated by

```
if n < 0
   parity = 0;
elseif rem(n,2) == 0
   parity = 2;
else
   parity = 1;
end
```

In two-way branching the `elseif` portion would, of course, be omitted.

## 6.4 Relations

The relational operators in MATLAB are

| | |
|---|---|
| < | less than |
| > | greater than |
| <= | less than or equal |
| >= | greater than or equal |
| == | equal |
| ~= | not equal |

Note that = is used in an assignment statement

whereas == is used in a relation. Relations may be connected or quantified by the logical operators

| | |
|---|---|
| & | and |
| | | or |
| ~ | not |

When applied to scalars, a relation is actually the scalar 1 or 0 depending on whether the relation is true or false. Try entering `3 < 5, 3 > 5, 3 == 5, and 3 == 3`. When applied to matrices of the same size, a relation is a matrix of 0's and 1's giving the value of the relation between corresponding entries. Try `a = rand(5), b = triu(a), a == b.`

A relation between matrices is interpreted by `while` and `if` to be true if each entry of the relation matrix is nonzero. Hence, if you wish to execute *statement* when matrices $A$ and $B$ are equal you could type

```
if A == B
    statement
end
```

but if you wish to execute *statement* when $A$ and $B$ are not equal, you would type

```
if any(any(A ~= B))
    statement
end
```

or, more simply,

```
if A == B else
    statement
end
```

Note that the seemingly obvious

```
if A ~= B, statement, end
```

will not give what is intended because *statement* would execute only if *each* of the corresponding entries of $A$ and $B$ differ. The functions any and all can be creatively used to reduce matrix relations to vectors or scalars. Two any's are required above because any is a vector operator (see section 7.2).

# 7. MATLAB Functions

## 7.1 Scalar Functions

Certain MATLAB functions operate essentially on scalars, but operate element-wise when applied to a matrix. The most common such functions are:

| | |
|---|---|
| abs | acos |
| round | atan |
| sin | rem (remainder) |
| cos | ceil |
| tan | exp |
| sqrt | log (natural log) |
| floor | log10 (log base 10) |
| asin | sign |

## 7.2 Vector Functions

Other MATLAB functions operate essentially on a vector (row or column), but act on an $m$-by-$n$ matrix ($m > 2$) in a column-by-column fashion to produce a row vector containing the results of their application to each column. Row-by-row action can be obtained by using the transpose – for example, `mean(A')'` – or by specifying the dimension along which to operate – for example, `mean(A,2)`. A few of these functions are:

| | | | |
|---|---|---|---|
| `max` | `sum` | `median` | `any` |
| `min` | `prod` | `mean` | `all` |
| `sort` | `std` | | |

For example, the maximum entry in a matrix A is given by `max(max(A))` rather than `max(A)`. Try it.

## 7.3 Matrix Functions

Much of MATLAB's power comes from its matrix functions. The most useful ones are:

| | |
|---|---|
| `eig` | eigenvalues and eigenvectors |
| `chol` | Cholesky factorization |
| `svd` | singular value decomposition |
| `inv` | inverse |
| `lu` | LU factorization |
| `qr` | QR factorization |
| `hess` | Hessenberg form |
| `schur` | Schur decomposition |
| `rref` | reduced row echelon form |
| `expm` | matrix exponential |
| `sqrtm` | matrix square root |
| `poly` | characteristic polynomial |
| `det` | determinant |
| `size` | size |
| `norm` | 1–norm, 2–norm, Frobenius–norm, ∞–norm |

| | |
|---|---|
| `cond` | condition number in the 2–norm |
| `rank` | rank |

MATLAB functions may have single or multiple output arguments. For example,

`y = eig(A)`, or simply `eig(A)`

produces a column vector containing the eigenvalues of $A$ whereas

```
[U,D] = eig(A)
```

produces a matrix $U$ whose columns are the eigenvectors of $A$ and a diagonal matrix $D$ with the eigenvalues of $A$ on its diagonal. Try it.

## 8. Command Line Editing And Recall

The command line in MATLAB can be easily edited. The cursor can be positioned with the left and right arrows and the **Backspace** (or **Delete**) key used to delete the character to the left of the cursor. Other editing features are also available. On a PC try the **Home**, **End**, and **Delete** keys; on a Unix system, PC, or Macintosh, the Emacs commands **Ctrl–a**, **Ctrl–e**, **Ctrl–d**, and **Ctrl–k** work; on other systems see `help cedit` or type `cedit`.

A convenient feature is use of the up and down arrows to scroll through the stack of previous commands. You can, therefore, recall a previous command line, edit it, and execute the revised line. For small routines, this is much more convenient than using an M-file, which requires moving between MATLAB and the editor (see section 12.1). For example, flopcounts (see section 13.1) for computing the inverse of matrices of various sizes can be compared by repeatedly recalling, editing, and executing

```
a = rand(8); flops(0), inv(a); flops
```

To compare plots of the functions $y = \sin(mx)$ and $y = \sin(nx)$ on the interval $[0, 2\pi]$ for various $m$ and $n$, you can do the same for the command line (place all on *one* line):

```
m=2; n=3; x=0:.01:2*pi; y=sin(m*x);
z=sin(n*x); plot(x,y,x,z)
```

# 9. Submatrices and Colon Notation

Vectors and submatrices are often used in MATLAB to achieve fairly complex data manipulation effects. "Colon notation" (which is used to both generate vectors and reference submatrices) and subscripting by integral vectors are keys to efficient manipulation of these objects. Creative use of these features to factorize operations lets you minimize the use of loops (which slows MATLAB) and to make code simple and readable. *Special effort should be made to become familiar with them.*

## 9.1 Generating Vectors

The expression `1:5` (met earlier in `for` statements) is actually the row vector `[1 2 3 4 5]`. The numbers need not be integers nor the increment one. For example,

```
0.2:0.2:1.2
```

gives `[0.2, 0.4, 0.6, 0.8, 1.0, 1.2]`, and

`5:-1:1` gives `[5 4 3 2 1]`

The following statements, for example, will generate a table of sines. Try it.

```
x  = [0.0:0.1:2.0]';
y  = sin(x);
[x y]
```

Note that because `sin` operates entry-wise, it produces a vector *y* from the vector *x*.

## 9.2 Accessing Submatrices

The colon notation can be used to access submatrices of a matrix. For example,

`A(1:4,3)` is the column vector consisting of the first four entries of the third column of *A*.

A colon by itself denotes an entire row or column:

`A(:,3)` is the third column of *A*, and
`A(1:4,:)` is the first four rows.

Arbitrary integral vectors can be used as subscripts:

`A(:,[2 4])` contains as columns, columns 2 and 4 of *A*.

Such subscripting can be used on both sides of an assignment statement:

`A (:,[2 4 5]) = B(:,1:3)` replaces columns 2,4,5 of *A* with the first three columns of *B*.

Note that the *entire* altered matrix *A* is displayed and assigned. Try it.

Columns 2 and 4 of A can be multiplied on the

right by the 2-by-2 matrix `[1 2;3 4]`:

```
A(:,[2,4]) = A(:,[2,4])*[1 2;3 4]
```

Once again, the entire altered matrix is displayed and assigned.

In subscripting, `end` denotes the index of the last element. If $x$ is an $n$-vector, what is the effect of the statement `x = x(end:-1:1)`? Try it.

Also try
`y = fliplr(x)` and `y = flipud(x')`.

To appreciate the usefulness of these features, compare these MATLAB statements with a Pascal, FORTRAN, or C routine to effect the same.

## 10. M-files

MATLAB can execute a sequence of statements stored in diskfiles. Such files are called "M-files" because they must have the file type `.m` as the last part of their filename. Much of your work with MATLAB will be in creating and refining M-files. M-files are usually created using your local editor.

There are two types of M-files: *script* files and *function* files.

### 10.1 Script Files

A *script* file consists of a sequence of normal MATLAB statements. If the file has the filename, say, `reflect.m`, then the MATLAB command `reflect` causes the statements in the file to be executed. Variables in a *script* file are global and will change the value of variables of the same name in the environment of the current MATLAB session.

*Script* files may be used to enter data into a large matrix; in such a file, entry errors can be easily corrected. If, for example, one enters in a diskfile `data.m`

```
A = [
1 2 3 4
5 6 7 8
];
```

then the MATLAB statement `data` causes the assignment given in `data.m` to be carried out. However, it is usually easier to use the MATLAB function `load` (see section 4.5).

An M-file can reference other M-files, including referencing itself recursively.

## 10.2 Function Files

*Function* files provide extensibility to MATLAB. You can create new functions specific to your problem, which will then have the same status as other MATLAB functions. Variables in a *function* file are by default local. A variable can, however, be declared *global* (see `help global`).

A *function* file is illustrated by the following simple example:

```
function y = randint(m,n)
% RANDINT Randomly generated integral
% matrix
% randint(m,n) returns an m-by-n
```

```
% matrix with entries between 0 and 9.
y = floor(10*rand(m,n));
```

A more general version of this function is the following:

```
function y = randint(m,n,a,b)
% RANDINT Randomly generated integral
% matrix
% randint(m,n) returns an m-by-n
% matrix with entries between 0 and 9.
% randint(m,n,a,b) returns entries
% between integers a and b.
if nargin < 3, a = 0; b = 9; end
y = floor((b-a+1)*rand(m,n)) + a;
```

This should be placed in a diskfile with filename `randint.m` (corresponding to the function name). The first line declares the function name, input arguments, and output arguments; without this line the file would be a *script* file. Then a MATLAB statement `z = randint(4,5)`, for example, causes the numbers 4 and 5 to be passed to the variables *m* and *n* in the function file, and the output result to be passed out to the variable *z*. Since variables in a *function* file are *local*, their names are independent of those in the current MATLAB environment.

Note that `nargin` ("number of input arguments") lets you set a default value of an omitted input variable, such as *a* and *b* in the example.

## 10.3 Multiple Output Variables

A function may also have multiple output arguments. For example,

```
function [mean, stdev] = stat(x)
% STAT Mean and standard deviation
%  For a vector x, stat(x) returns
%  the mean of x.
%  [mean, stdev] = stat(x) returns both
%  the mean and standard deviation
%  (normalized by N, producing the
%  second moment of the sample
%  about its mean).
%  For a matrix x, stat(x) acts
% columnwise.
[m n] = size(x);
if m == 1
   m = n; % handle case of a row vector
end
mean = sum(x)/m;
stdev = sqrt(sum(x.^2)/m – mean.^2);
```

Once this is placed in a diskfile `stat.m`, a MATLAB command `[xm, xd] = stat (x)`, for example, assigns the mean and standard deviation of the entries in the vector $x$ to $xm$ and $xd$, respectively. Single assignments can also be made with a function having multiple output arguments. For example, `xm = stat(x)` (no brackets needed around $xm$) will assign the mean of $x$ to $xm$.

## 10.4 Comments, Documentation for Help

The `%` symbol indicates that the rest of the line is a comment; MATLAB will ignore the rest of the line. Moreover, the first few contiguous comment lines are used to document the M-file. They are available to the on-line help facility and will be displayed if, for example, `help stat` is entered. Such documentation should *always* be included in a function file.

## 10.5 Producing Efficient Code

The function `stat.m` illustrates some of the MATLAB features that can be used to produce efficient code. Note, for example, that `x.^2` is the matrix of squares of the entries of $x$, that sum is a vector function (section 7.2), that `sqrt` is a scalar function (section 7.1), and that the division in `sum(x)/m` is a matrix-scalar operation. Thus all operations are vectorized and loops avoided.

If you can't vectorize some computations, you can make your `for` loops go faster by preallocating any vectors or matrices in which output is stored. For example, by including the second statement below, which uses the function `zeros`, space for storing $E$ in memory is preallocated. Without this, MATLAB must resize $E$ one column larger in each iteration, slowing execution.

```
M = magic(6);
E = zeros(6,50);
for j = 1:50
   E(:,j) = eig(M^j);
end
```

## 10.6 Advanced Features

Some more advanced features are illustrated by the following function. As noted earlier, some of the input arguments of a function, such as `tol` in this example, may be made optional through use of `nargin` ("number of input arguments"). The variable `nargout` can be similarly used. Note that the fact that a relation is a number (1 when true; 0 when false) is used and that, when `while` or `if` evaluates a relation, "nonzero" means "true" and 0 means "false". Finally, the MATLAB function `feval` permits a string naming another function to be an input variable. (Also see `eval`.)

```
function [b, steps] = bisect(fun,x,tol)
% BISECT Zero of a function of one
% variable via the bisection method.
% bisect(fun,x) returns a zero of the
% function.
% fun is a string containing the name
% of a real-valued MATLAB function of
% a single real variable; ordinarily
% functions are defined in M-files.
% x is a starting guess.
% The value returned is near a point
% where fun changes sign.
% For example, bisect('sin',3) is pi.
% Note the quotes around sin.
%
% An optional third input argument
% sets a tolerance for the relative
% accuracy of the result.
% The default is eps.
```

```
% An optional second output argument
% gives a matrix containing a trace
% of the steps;
% the rows are of form [c f(c)].

% Initialization
if nargin < 3, tol = eps; end
trace =(nargout == 2);
if x ~=0, dx=x/20; else, dx=1/20; end
a = x - dx; fa = feval(fun,a);
b = x + dx; fb = feval(fun,b);

% Find change of sign.
while (fa > 0) == (fb > 0)
  dx = 2.0*dx;
  a = x - dx; fa = feval(fun,a);
  if (fa > 0) ~= (fb > 0), break, end
  b = x + dx; fb = feval(fun,b);end
if trace, steps = [a fa; b fb];
end

% Main loop
while abs(b-a) > 2.0*tol*max(abs(b),1.0)
  c = a + 0.5*(b-a); fc = feval(fun,c);
  if trace, steps=[steps; [c fc]]; end
  if (fb > 0) == (fc > 0)
    b = c; fb = fc;
  else
    a = c; fa = fc;
  end
end
```

Some of MATLAB's functions are built-in; others

are distributed as M-files. The actual listing of any non-built-in M-file, MATLAB's or your own, can be viewed with the MATLAB command `type` *functionname*. Try entering `type eig`, `type vander`, and `type rank`.

## 10.7 Calling Priorities, Subfunctions, Private Functions

When MATLAB comes upon a new name, it resolves it into a specific function by following these steps:

1. Checks to see if the name is a variable.
2. Checks to see if the name is a subfunction (a MATLAB function that resides in the same M-file as the calling function).
3. Checks to see if the name is a private function (a MATLAB function that resides in a directory named `private`, which is accessible only to M-files in the directory immediately above it).
4. Checks to see if the name is a function on the MATLAB search path (see `help path`). MATLAB uses the first file it encounters with the specified name.

# 11. Strings, Error Messages, Input

Text strings are entered into MATLAB surrounded by single quotes. For example,

```
s = 'This is a test'
```

assigns the given text string to the variable *s*.

Text strings and numeric matrices can be displayed with the function `disp` and `disp(magic(3))`. For example,

```
disp ('this message is hereby displayed')
```

## 11.1 Error Messages

Error messages are best displayed with the function `error`

```
error ('Sorry, the matrix must be symmetric')
```

because when placed in an M-file, `error` aborts execution of the M-file.

## 11.2 Input

In an M-file the user can be prompted to interactively enter input data with the function input. When, for example, the statement

```
iter = input ('Enter the number of iterations:')
```

is encountered, the prompt message is displayed and execution pauses while the user keys in the input data. Upon pressing the **Return** key, the data is assigned to the variable `iter` and execution resumes.

# 12. Managing M-files

While using MATLAB you will often want to create or edit an M-file with the local editor and then return to MATLAB. You want to keep MATLAB active while editing a file; otherwise, all variables would be lost upon exiting.

## 12.1 Executing System Commands (!–Feature)

This can be easily done using the !–feature. If, while in MATLAB, you precede it with an !, any system command, such as those for editing, printing, or copying a file, can be executed without exiting MATLAB. If, for example, the system command `ed` accesses your editor, the MATLAB command

```
>> !ed reflect.m
```

lets you edit the file named `reflect.m` using your local editor. Upon leaving the editor, you will be returned to MATLAB just where you left it. The Macintosh and PC versions have their own editor and any file on the path may be edited by typing `edit` *filename* from the command line. For example, try `edit hilb`.

However, as noted earlier, on systems permitting multiple processes, such as one running Unix or MS Windows, it may be preferable to keep both MATLAB and your local editor active, keeping one process suspended while working in the other. If these processes can be run in multiple windows, you will want to keep MATLAB active in one window and your editor active in another.

You should consult your instructor or your local computing center for details of the local installation.

## 12.2 Working with Directories and Files

When you are in MATLAB, the command `pwd` returns the name of the present working directory and `cd` will change the working directory. The command `dir` (or `ls`) lists the contents of the working directory, whereas the command `what` lists only the MATLAB specific files in the directory, grouped by file type. The MATLAB commands `delete` and `type` can be used to delete a diskfile and print an M-file to the screen, respectively. Although these commands may duplicate system commands, they avoid the use of an `!`. You may enjoy entering the command `why` a few times.

## 12.3 MATLAB and `path`

M-files must be in a directory accessible to MATLAB. M-files in the present working directory are always accessible. On most mainframe or workstation network installations, personal M-files that are stored in a subdirectory of your home directory named `matlab` will be accessible to MATLAB from any directory in which you are working. The current list of directories in MATLAB's search path is obtained by the command `path`. This command can also be used to add or delete directories from the search path. See `help path`. The command `which` locates functions and files on the path. For example, type `which hilb`. The Macintosh and PC have interactive browsers (try `pathtool`).

## 12.4 Debugging

Many debugging tools are available. See the list of functions given in section 19.4. The Macintosh and PC have interactive debuggers built into the MATLAB editor.

# 13. Comparing Efficiency of Code

Two measures of the efficiency of an algorithm are the number of floating point operations (`flops`) performed and the elapsed time.

## 13.1 Flops

The MATLAB function `flops` keeps a running total of the floating point operations performed. The command `flops(0)` (not `flops = 0!`) resets flops to 0. Hence, entering `flops(0)` immediately before executing an algorithm and `flops` immediately after gives the flop count for the algorithm. For example, the number of flops required to solve a given linear system via Gaussian elimination can be obtained with:

```
flops(0), x = A\b; flops
```

## 13.2 Elapsed Time (Tic, Toc)

The elapsed time (in seconds) can be obtained with the stopwatch timers `tic` and `toc`; `tic` starts the timer and `toc` returns the elapsed time. Hence, the commands

`tic,` *any statement*`, toc`

will return the elapsed time for execution of the statement. The elapsed time for solving the linear system above can be obtained, for example, with

```
tic, x = A\b; toc
```

You may wish to compare this time, and flop count, with that for solving the system using `x = inv(A)*b;`. Try it.

It should be noted that, on timesharing machines, elapsed time may not be a reliable measure of the efficiency of an algorithm because the rate of execution depends on how busy the machine is at the time.

### 13.3 Profile

MATLAB provides an M-file profiler that lets you see how much computation time is used by each line of an M-file. The command to use is `profile` (see `help profile` for details).

## 14. Output Format

Although all computations in MATLAB are performed in double precision, the format of the displayed output can be controlled by the following commands.

| | |
|---|---|
| `format short` | fixed point, 4 decimal places |
| `format long` | fixed point, 14 decimal places |
| `format short e` | scientific notation, 4 dec. places |
| `format long e` | scientific notation, 15 dec. places |
| `format rat` | approx. by ratio of small integers |
| `format hex` | hexadecimal format |
| `format bank` | fixed dollars and cents (imaginary part not displayed) |
| `format+` | +,–, blank |

`format short` is the default. Once invoked, the chosen format remains in effect until changed. Note that these commands only modify the display, not the precision of the number.

The command `format compact` suppresses most blank lines allowing more information to be placed on the screen or page. The command `format loose` returns to the non-compact format. These two commands are independent of the other format commands.

# 15. Graphics

MATLAB can produce planar plots of curves, three-dimensional plots of curves, three-dimensional mesh surface plots, and three-dimensional faceted surface plots. The primary commands for these facilities are `plot`, `plot3`, `mesh`, `surf`, and `light`. An introduction to each of these is given below.

To preview some of these capabilities, enter the command `demo` and select some of the visualization and graphics demos.

## 15.1 Planar Plots (`plot`)

The `plot` command creates linear x–y plots; if *x* and *y* are vectors of the same length, the command `plot(x,y)` opens a graphics window and draws an x–y plot of the elements of *x* versus the elements of *y*. You can, for example, draw the graph of the sine function over the interval –4 to 4 with the following commands:

```
x = -4:0.01:4; y = sin(x); plot(x,y)
```

Try it. The vector $x$ is a partition of the domain with meshsize 0.01, and $y$ is a vector giving the values of sine at the nodes of this partition (recall that `sin` operates entry-wise).

When plotting a curve, the `plot` routine is actually connecting consecutive points induced by the partition with line segments. Thus, the meshsize should be chosen sufficiently small to render the appearance of a smooth curve.

You will usually want to keep the current graphics window ("figure") exposed, but moved to the side, and the command window active.

As a second example, you can draw the graph of $y = e^{-x^2}$ over the interval –1.5 to 1.5 as follows:

```
x = -1.5:.01:1.5; y = exp(-x.^2);plot(x,y)
```

Note that one must precede ^ by a period (.) to ensure that it operates entry-wise.

The function `zoom` lets you zoom in and out on a two-dimensional plot. See `help zoom`.

## 15.2 Multiple Figures

You can have several concurrent graphics figures, one of which will at any time be the designated "current" figure in which graphs from subsequent plotting commands will be placed. If, for example, figure 1 is the current figure, then the command `figure(2)` (or simply `figure`) will open a second figure (if necessary) and make it the current figure. The command `figure(1)` will then expose figure 1 and make it again the current figure. The command `gcf` will return the number of the current figure.

## 15.3 Graph of a Function (fplot)

MATLAB supplies a function fplot to easily and efficiently plot the graph of a function. For example, to plot the graph of the function above, you can first define the function in an M-file called, say, expnormal.m containing

```
function y = expnormal(x)
y = exp(-x.^2);
```

Then the command

```
fplot('expnormal',[-1.5,1.5])
```

will produce the graph over the indicated x-domain. Try it.

A quicker way to see the same result would be

```
fplot('exp(-i^2)',[-1.5,1.5])
```

## 15.4 Parametrically Defined Curves

Plots of parametrically defined curves can also be made. Try, for example (on one line),

```
t = 0:.001:2*pi; x = cos(3*t);
   y = sin(2*t); plot(x,y)
```

## 15.5 Titles, Labels, Text in a Graph

The graphs can be given titles, axes labeled, and text placed within the graph with the following commands, which take a string as an argument.

| | |
|---|---|
| title | graph title |
| xlabel | x-axis label |
| ylabel | y-axis label |
| gtext | place text on graph using the mouse |

| | |
|---|---|
| `text` | position text at specified coordinates |

For example, the command

```
title ('Best Least Squares Fit')
```

gives a graph a title. The command `gtext('The Spot')` lets you interactively place the designated text on the current graph by placing the mouse cross hair at the desired position and clicking the mouse. To place text in a graph at designated coordinates, use the command `text` (see `help text`).

The command `grid` will place grid lines on the current graph.

## 15.6 Control of Axes and Scaling (axis)

By default, the axes are auto-scaled. This can be overridden by the command `axis`. Some features of axis are:

| | |
|---|---|
| `axis` | $([x_{min}, x_{max}, y_{min}, y_{max}])$ sets axis scaling to prescribed limits |
| `axis (axis)` | freezes scaling for new plots |
| `axis auto` | returns to auto-scaling |
| `v = axis` | vector $v$ shows current scaling |
| `axis square` | axes same size (but not scale) |
| `axis equal` | same scale and tic marks on axes |
| `axis off` | removes the axes |
| `axis on` | restores the axes |

The `axis` command should be given after the `plot` command.

## 15.7 Multiple Plots

Two ways to make multiple plots on a single graph are illustrated by

```
X = 0:.01:2*pi;
y1=sin(x); y2=sin(2*x);y3=sin(4*x);
plot(x,y1,x,y2,x,y3)
```

and by forming a matrix Y containing the functional values as columns

```
x = 0:.01:2*pi;
Y = [sin(x)', sin(2*x)', sin(4*x)']
plot(x,Y)
```

Another way is with `hold`. The command `hold on` freezes the current graphics screen so that subsequent plots are superimposed on it. The axes may, however, become rescaled. Entering `hold off` releases the "hold".

The function `legend` places a legend in the current figure to identify the different graphs. See `help legend`.

## 15.8 Line Types, Marker Types, Colors

You can override the default line types, marker types and colors. For example,

```
x = 0:.01:2*pi;
y1 = sin(x); y2 = sin(2*x);
  y3 = sin(4*x);
plot(x,y1,'—',x,y2,':',x,y3,'+')
```

renders a dashed line and dotted line for the first two graphs, whereas for the third the symbol + is placed at each node. The line and marker types are:

```
Line types: solid(-), dashed(—),
  dotted(:), dashdot(-.)

Marker types: point(.) plus(+), star(*),
  circle(o), x-mark(x),
  square(s),diamond(d), triangle-down(v),
  triangle-up(^),
  triangle-left(<),triangle-right(>),
  pentagram(p), hexagram(h)
```

Colors can be specified for the line and marker types.

```
Colors:yellow(y), magenta (m),
  cyan(c), red(r), green(g), blue(b),
  white(w), black(k)
```

For example, `plot (x,y,'r—')` plots a red dashed line.

## 15.9 Subplot, Specialized Plots

The command `subplot` partitions a figure so that several small plots can be placed in one figure. See `help subplot`.

Other specialized planar plotting functions you may wish to explore via help are:

```
polar, bar, hist, quiver, compass,
feather, rose, stairs, fill
```

## 15.10 Graphics Hardcopy (`print`)

A hardcopy of the current graphics figure is most easily obtained with the MATLAB command `print`. Entered by itself, it sends a high-resolution copy of the current graphics figure to the default printer.

The `printopt` M-file is used to specify the default setting used by the `print` command. If desired, you can change the defaults by editing this file (see `help printopt`).

The command `print` *filename* saves the current graphics figure to the designated filename in the default file format. If, for example, PostScript is the default file format, then

```
print lissajous
```

creates a PostScript file `lissajous.ps` of the current graphics figure, which can subsequently be printed using the system `print` command. If the filename already exists, it will be overwritten unless you use the `append` option. The command

```
print -append lissajous
```

appends the (hopefully different) current graphics figure to the existing file `lissajous.ps`. In this way you can save several graphics figures in a single file.

The default settings can, of course, be overwritten. For example,

```
print -deps -f3 saddle
```

saves to an Encapsulated PostScript file `saddle.eps` the graphics figure 3, even if it is not the current figure. The options "`-d`" and "`-f`" denote "device" and "figure", respectively.

Saving a figure for later reloading can be done by creating an M-file which will generate the figure. The command

```
print -dmfile saddle
```

creates an M-file containing the necessary information to later reproduce the figure.

## 15.11 Three-Dimensional Curve Plots (plot3)

Completely analogous to plot in two dimensions, the command plot3 produces curves in three-dimensional space. If $x$, $y$, and $z$ are three vectors of the same size, then the command plot3(x,y,z) produces a perspective plot of the piecewise linear curve in 3-space passing through the points whose coordinates are the respective elements of $x$, $y$, and $z$. These vectors are usually defined parametrically. For example,

```
t = .01:.01:20*pi;
x = cos(t); y = sin(t); z = t.^3;
plot3(x,y,z)
```

produces a helix that is compressed near the x–y plane (a "slinky"). Try it.

Just as for planar plots, a title and axis labels (including zlabel) can be added. The features of the axis command described there also hold for three-dimensional plots; setting the axis scaling to prescribed limits will, of course, now require a 6-vector.

## 15.12 Mesh and Surface Plots (mesh, surf)

Three-dimensional wire mesh surface plots are drawn with the command mesh. The command mesh(z) creates a three-dimensional perspective plot of the elements of the matrix $z$. The mesh surface is defined by the z-coordinates of points above a rectangular grid in the x–y plane. Try `mesh(eye(20))`.

Similarly, three dimensional faceted surface plots are drawn with the command `surf`.
Try `surf(eye(20))`.

To draw the graph of a function $z=f(x, y)$ over a rectangle, first define vectors *xx* and *yy*, which give partitions of the sides of the rectangle. The function `meshgrid` then creates a matrix *x*, each row of which equals *xx* (whose column length is the length of *yy*) and similarly a matrix *y*, each column of which equals *yy*, as follows:

```
[x,y] = meshgrid(xx,yy);
```

A matrix *z*, to which `mesh` or `surf` can be applied, is then computed by evaluating *f* entry-wise over the matrices *x* and *y*.

You can, for example, draw the graph of

$z=e^{-x^2-y^2}$ over the square `[-2, 2]x[-2, 2]` as follows (try it):

```
xx = -2:.2:2;
yy = xx;
[x,y] = meshgrid(xx,yy);
z = exp(-x.^2 - y.^2);
mesh(z)
```

You could, of course, replace the first three lines of the preceding with

```
[x,y] = meshgrid(-2:.2:2, -2:.2:2);
```

Try this plot with `surf` instead of `mesh`.

Note that you must use x.^2 and y. –^2 instead

of x-^2 and y-^2 to ensure that the function acts entry-wise on *x* and *y*.

As noted earlier, the features of the `axis` command described in section 15.6 also hold for three-dimensional plots, as do the commands for titles, axis labeling and the command `hold`.

## 15.13 Color Shading and Color Profile

The color shading of surfaces is set by the `shading` command. There are three settings for shading: `faceted` (default), `interpolated`, and `flat`. These are set by the commands

```
shading faceted, shading interp, and
shading flat
```

Note that on surfaces produced by `surf`, the settings `interpolated` and `flat` remove the superimposed mesh lines. Experiment with various shadings on the surface produced above.

The command `shading` (as well as `colormap` and `view` below) should be entered after the `surf` command.

The color profile of a surface is controlled by the `colormap` command. Available predefined color maps include:

```
hsv (default), hot, cool, jet, pink, copper,
flag, gray, bone, prism, white
```

The command `colormap(cool)`, for example, sets a certain color profile for the current figure. Experiment with various color maps on the surface produced above. See also `help colorbar`.

## 15.14 Perspective of View (`view`, `light`, `rotate3d`, and camera)

The command `view` can be used to specify in spherical or Cartesian coordinates the viewpoint from which the three-dimensional object is to be viewed. See `help view`.

The command `rotate3d` can be used to interactively set the view of a three-dimensional plot, using the mouse. In the example below, click on the surface, and move the mouse.

The MATLAB function `peaks` generates an interesting surface on which to experiment with `shading`, `colormap`, and `view`. See `help peaks`.

Try

```
peaks;
rotate3don;
```

In MATLAB 5, light sources and camera position can be set. Taking the peaks surface from the example above, type `light` to add a light source. See *Using MATLAB Graphics*, for camera help.

## 15.15 Parametrically Defined Surfaces

Plots of parametrically defined surfaces can also be made. The MATLAB functions `sphere` and `cylinder` generate such plots of the named surfaces. (See `type sphere` and `type cylinder`.) The following is an example of a similar function that generates a plot of a torus by utilizing spherical coordinates.

```
function [x,y,z] = torus(r,n,a)
% TORUS Generate a torus
% torus(r,n,a) generates a plot of a
% torus with central radius a and
% lateral radius r.
% n controls the number of facets
% on the surface.
% These input variables are optional
% with defaults r = 0.5, n = 30, a = 1.
%
% [x,y,z] = torus(r,n,a) generates
% three (n+1)-by-(n+1) matrices so that
% surf(x,y,z) will produce the torus.
%
% See also SPHERE, CYLINDER
  if nargin < 3, a = 1; end
  if nargin < 2, n = 30; end
  if nargin < 1, r = 0.5; end
  theta = pi*(0:2:2*n)/n;
  phi = 2*pi*(0:2:n)'/n;
  xx=(a + r*cos(phi))*cos(theta);
  yy=(a + r*cos(phi))*sin(theta);
  zz=r*sin(phi)*ones(size(theta));
  if nargout == 0
     surf(xx,yy,zz)
     ar = (a + r)/sqrt(2);
     axis([-ar,ar,-ar,ar,-ar,ar])
  else
      x = xx; y = yy; z = zz;
  end
```

Other three-dimensional plotting functions you may wish to explore via `help` are

```
meshz, surfc, surfl, contour, pcolor
```

# 16. Advanced Graphics

MATLAB possesses a number of other advanced graphics capabilities. Significant ones are object based graphics, called Handle Graphics,® and Graphical User Interface (GUI) tools.

## 16.1 Handle Graphics

Beyond those just described, MATLAB's graphics system provides low-level functions that let you control virtually all aspects of the graphics environment to produce sophisticated plots. The commands `set` and `get` allow access to all the properties of your plots. Try `set(gcf)` to see some of the properties of a figure which you can control. This system is called Handle Graphics; see the *Using MATLAB Graphics* for more information.

## 16.2 Graphical User Interface (GUI)

MATLAB's graphics system also provides the ability to add sliders, push-buttons, menus and other user interface controls to your own figures. For information on creating user interface controls, try `help uicontrol`. This allows you to create interactive graphical based applications. For more information, see the *Building GUIs with MATLAB*.

A new capability in MATLAB 5 is the GUI

Development Environment, Guide, which can be used to create and lay out GUI's. Try `guide`.

# 17. Sparse Matrix Computations

In performing matrix computations, MATLAB normally assumes that a matrix is dense; that is, any entry in a matrix *may* be nonzero. If, however, a matrix contains enough zero entries, computation time could be reduced by avoiding arithmetic operations on zero entries, and less memory could be required by storing only the nonzero entries of the matrix. This increase of efficiency in time and storage can make feasible the solution of significantly larger problems than would otherwise be possible. MATLAB provides the capability to take advantage of the sparsity of matrices.

## 17.1 Storage Modes (full, sparse)

MATLAB has two storage modes, full and sparse, with full the default. The functions `full` and `sparse` convert between the two modes. For a matrix A, full or sparse, `nnz(A)` returns the number of nonzero elements in A.

A sparse matrix is stored as a linear array of its nonzero elements along with their row and column indices. If a full tridiagonal matrix F is created via, say,

```
F=floor(10*rand(6));
F=triu(tril(F,1),-1);
```

then the statement `S = sparse(F)` will convert *F* to sparse mode. Try it. Note that the output lists the nonzero entries in column major order along with their row and column indices. The statement `F = full(S)` restores $S$ to full storage mode. You can check the storage mode of a matrix A with the command `issparse(A)`.

## 17.2 Generating Sparse Matrices

A sparse matrix is, of course, usually generated directly rather than by applying the function `sparse` to a full matrix. A sparse banded matrix can be easily created via the function `spdiags` by specifying diagonals. For example, a familiar sparse tridiagonal matrix is created by

```
m = 6; n = 6; e = ones(n,1); d = -2*e;
T = spdiags([e,d,e],[-1,0,1],m,n)
```

Try it. The integral vector `[-1, 0, 1]` specifies in which diagonals the columns of `[e, d, e]` should be placed (use `full(T)` to see the full matrix T and `spy(T)` to view T graphically). Experiment with other values of $m$ and $n$ and, say,
`[-3, 0, 2]` instead of `[-1, 0, 1]`. See `help spdiags` for further features of `spdiags`.

The sparse analogs of eye, zeros, ones, and randn for full matrices are, respectively,

```
speye, sparse, spones, sprandn
```

The latter two take a matrix argument and replace only the nonzero entries with ones and normally distributed random numbers, respectively. sprandn also permits the sparsity structure to be randomized. The command sparse(m,n) creates a sparse zero matrix.

The versatile function sparse also permits creation of a sparse matrix via listing its nonzero entries. Try, for example,

```
i = [1 2 3 4 4 4]; j = [1 2 3 1 2 3];
s = [5 6 7 8 9 10];
S = sparse(i,j,s,4,3), full(S)
```

Note that if there are repeated entries in a sparse matrix, then the entries are added together. For example, the commands below create a matrix whose diagonals are 2, 1, and 1.

```
i = [1,2,3,1]; j = [1,2,3,1];
s = [1,1,1,1]; S = sparse(i,j,s);
full(S)
```

In general, if the vector $s$ lists the nonzero entries of $S$ and the integral vectors $i$ and $j$ list their corresponding row and column indices, then

```
sparse(i,j,s,m,n)
```

will create the desired sparse *m–by–n* matrix $S$. As another example try

```
n = 6; e = floor(10*rand(n-1,1));
E = sparse(2:n,1:n-1,e,n,n)
```

## 17.3 Computation with Sparse Matrices

The arithmetic operations and most MATLAB functions can be applied independent of storage mode. The storage mode of the result? Operations on full matrices always give full results. Selected other results are (S=sparse, F=full):

```
Sparse:      S+S, S*S, S.*S, S.*F,
             S^n, S.^n, S\s
```

```
Sparse:      inv(S), chol(S), lu(S),
             diag(S), max(S), sum(S)
Full:        S+F, S*F, S\F, F\S
```

**Remark**: If $F$ is a scalar, then `S*F`, `S\F`, and `F\S` will remain sparse data structures.

To compute the eigenvalues or singular values of a sparse matrix $S$, you must convert $S$ to a full matrix and then use `eig` or `svd`: `eig(full(S))` or `svd(full(S))`. A matrix built from blocks, such as `[A, B; C, D]`, is stored in sparse mode if any constituent block is sparse. If $S$ is a "large" sparse matrix and you wish only to compute some of the eigenvalues or singular values, then you can use the `eigs` or `svds` functions. These commands will accept sparse data structures: `eigs(S)`, `svd(S)`.

It is instructive to compare, for the two storage modes, the efficiency of solving a tridiagonal system of equations for, say, `n = 20, 50, 500, 1000` by entering, recalling and editing the following command lines (You will want to consolidate these into two or three lines).

```
n = 20; e = ones(n,1); d = -2*e;
T = spdiags([e,d,e],[-1,0,1],n,n);
A = full(T);
b = ones(n,1); s = sparse(b);
tic, for j = 1:10, T\s; end
sparsetime = toc/10
tic, for j = 1:10, A\b; end
fulltime = toc/10
```

# 18. The Symbolic Math Toolbox

The Symbolic Math Toolbox, which utilizes the Maple V kernel as its computer algebra engine, lets you perform symbolic computation from within MATLAB. Under this configuration, MATLAB's numeric and graphic environment is merged with Maple's symbolic computation capabilities. The toolbox M-files that access these symbolic capabilities have names and syntax that will be natural for the MATLAB user. A limited version of the Symbolic Math Toolbox, moreover, is included within the Student Edition of MATLAB.

Since the Symbolic Math Toolbox is not part of basic MATLAB, it may not be installed on your system, in which case this section will not apply.

## 18.1 Calculus (diff, int, limit, and taylor)

The function diff computes the symbolic derivative of a function defined by a symbolic expression. First, to define a symbolic expression, you should create symbolic variables and then proceed to build an expression as you would mathematically. For example,

```
syms x
f = x^2*exp(x);
diff(f)
```

creates a symbolic variable $x$, builds the symbolic expression $f = x^2 e^x$, and returns the symbolic derivative of $f$ with respect to $x$: 2*x*exp(x)+x^2*exp(x). Try it.

Next

```
syms t
diff(sin(pi*t))
```

returns the derivative of $\sin(\pi t)$, as a function of $t$.

Partial derivatives can also be computed. Try the following.

```
syms x y
g = x*y + x^2;
diff(g)   % computes ∂g/∂x
diff(g,x)   % also ∂g/∂x
diff(g,y)   % ∂g/∂y
```

To permit omission of the second argument for functions such as the above, MATLAB chooses a default symbolic variable for the symbolic expression. The function `findsym` expression will return MATLAB's choice. Its rule is, roughly, to choose that lower case letter, other than $i$ and $j$, nearest $x$ in the alphabet.

You can, of course, override the default choice as shown above. Try, for example,

```
syms x1 x2
F = x*(x1*x2 + x1 - 2);
diff(F)                  %∂F/∂x
diff(F,x1)               %∂F/∂x1
diff(F,x2)               %∂F/∂x2
G = cos(theta*x)
diff(G,theta)            %∂G/∂theta
```

The second derivative, for example, can be obtained by the command

```
diff(sin(2*x),x,2)
```

With a numeric argument, diff is the difference operator of basic MATLAB, which can be used to numerically approximate the derivative of a function. See help diff.

The function int attempts to compute the indefinite integral (antiderivative) of a function defined by a *symbolic* expression. Try, for example,

```
syms a b t x y z theta
int(sin(a*t + b))
int(sin(a*theta + b),theta)
int(x*y^2 + y*z,y)
int(x^2*sin(x))
```

Note that, as with diff, when the second argument of int is omitted, the default symbolic variable (as selected by findsym) is chosen as the variable of integration.

In some instances, int will be unable to give a result in terms of elementary functions. Consider, for example,

```
int(exp(-x^2))
int(sqrt(1+x^3))
```

In the first case the result is given in terms of the error function erf, whereas in the second, the integral is reduced via integration by parts to another integral.

The function pretty will "pretty-print" a symbolic expression in an easier-to-read form resembling typeset mathematics (but still not very "pretty").

Try, for example,

```
pretty(sqrt(1 + x^3))
int(sqrt(1 + x^3)); pretty(ans)
```

Definite integrals can also be computed by using additional input arguments. Try, for example,

```
int(sin(x),0,pi)
int(sin(theta),theta,0,pi)
```

In the first case, the default symbolic variable $x$ was used as the variable of integration to compute

$$\int_0^1 \sin(x)\,dx,$$

whereas in the second *theta* was chosen. Other definite integrals you can try are:

```
int(x^5,1,2)
int(log(x),1,4)
int(x*exp(x),0,2)
int(exp(-x^2),0,inf)
```

It is important to realize that the results returned are symbolic expressions, not numeric ones. The function `double` will convert these into MATLAB floating point numbers, if desired. For example, the result returned by the first integral above is

```
ans =
21/2
```

Entering `double(ans)` then returns the MATLAB numeric result

```
ans =
10.5000
```

Alternatively, you can use the function vpa (variable precision arithmetic; see section 18.2) to convert the expression into a symbolic number of arbitrary precision.

For example,

```
int(exp(-x^2),0,inf)
ans =
1/2*pi^(1/2)
vpa(ans,25)
ans =
.8862269254527580136490835
```

symbolically gives the result to 25 significant digits.

You may wish to contrast these techniques with the MATLAB numerical integration functions quad and quad8.

The `limit` function is used to compute the symbolic limits of various expressions. For example,

```
syms h n x
limit((1 + x/n)^n ,n,inf)
```

computes the limit of $(1 + x/n)^n$ as $n \rightarrow \infty$. You should also try

```
limit(sin(x),x,0)
limit((sin(x+h) - sin(x))/h,h,0)
```

The `taylor` function computes the Maclaurin and Taylor series of symbolic expressions. For example,

```
taylor(cos(x) + sin(x))
```

returns the 5th order Maclaurin polynomial

approximating cos(*x*) + sin(*x*). The command,

```
taylor(cos(x^2),8,x,pi)
```

returns the 8th degree Taylor approximation to $\cos(x^2)$ centered at the point $x_o = \pi$.

## 18.2 Variable Precision Arithmetic (vpa)

With this toolbox, three kinds of arithmetic operations are available.

**Numeric** MATLAB's floating point arithmetic

**Rational** Maple's exact symbolic arithmetic

**VPA** Maple's variable precision arithmetic

One can obtain exact rational results with, for example,

```
sym(13/17 + 17/23)
```

You are already familiar with numeric computations. For example, with `format long`, the following will give a numeric result.

```
pi*log(2)
ans
2.17758609030360
```

MATLAB's numeric computations are done in approximately 16 decimal digit floating point arithmetic.

With vpa, you can obtain results to arbitrary precision, within the limitations of time and memory. For example, try

```
vpa(pi*log(2))
vpa(pi*log(2),50)
```

Upon startup, the default precision for vpa is 32.

Hence, the first result is accurate to 32 digits, whereas the second is accurate to the specified 50 digits. The default precision can be changed with the function digits.

While the rational and VPA computations can be more accurate, they are in general slower than numeric computations.

## 18.3 Simplification (factor, expand, simplify)

Convenient algebraic manipulations of symbolic expressions are available.

The function expand distributes products over sums and applies other identities, whereas factor attempts to do the reverse. The function collect views a symbolic expression as a polynomial in its symbolic variable (which may be specified), and collects all terms with the same power of the variable. To explore these capabilities, try the following.

```
syms a b x y z
expand((a+b)^5), factor(ans)
expand(exp(x+y))
expand(sin(x+2*y)), factor(x^6-1)
collect(x*(x*(x+3)+5)+1), horner(ans)
collect((x+y+z)*(x-y-z))
collect((x+y+z)*(x-y-z),y)
collect((x+y+z)*(x-y-z),z)
diff(x^3*exp(x)), factor(ans)
```

The powerful function simplify applies many identities in an attempt to reduce a symbolic expression to a simple form.

Try, for example,

```
simplify(sin(x)^2+cos(x)^2)
simplify(exp(5*log(x) + 1))
d = diff((x^2 + 1)/(x^2 - 1))
simplify(d)
```

The alternate function `simple` computes several "simplifications" and chooses the shortest of them. It often gives better results on expressions involving trigonometric functions. Try the following commands

```
simplify(cos(x)+(-sin(x)^2)^(1/2))
simple(cos(x)+(-sin(x)^2)^(1/2))
simplify((1/x^3+6/x^2+12/x+8)^(1/3))
simple((1/x^3+6/x^2+12/x+8)^(1/3))
```

The function `subs` replaces all occurrences of the symbolic variable in a expression by a specified second expression. This corresponds to composition of two functions. Try, for example,

```
syms x s t
subs(sin(x),x,pi/3)
subs(g*t^2/2,t,sqrt(2*s))
subs(sqrt(1-x^2),x,cos(x))
subs(sqrt(1-x^2),1-x^2,cos(x))
```

The general idea is that the third argument (`NEW`) replaces the second argument (`OLD`) in the first argument (`EXPR`): `subs(expr,old,new)`.

The function `factor` can also be applied to an integer argument to compute the prime factorization of the integer.

Try, for example,

```
factor(sym('4248'))
factor(sym('4549319348693'))
factor(sym('4549319348597'))
```

## 18.4 Graphs of Functions (ezplot, funtool)

The MATLAB function fplot (see section 15.3) provides a tool to conveniently plot the graph of a function. Since it is, however, the name of the function to be plotted that is passed to fplot, the function must first be defined in an M-file (or else be a built-in function or inline function).

In the Symbolic Math Toolbox, ezplot lets you plot the graph of a function directly from its defining symbolic expression. For example, try

```
syms t x
ezplot(sin(2*x))
ezplot(t+3*sin(t))
ezplot(2*x/(x^2-1))
ezplot(1/(1+30*exp(-x))) % (a logistic
                         % curve)
```

By default, the *x*-domain is [-2*pi, 2*pi]. This can be overridden by a second input variable, as with ezplot(x*sin(1/x),[-.2 .2]).

You will often need to specify the *x*-domain to zoom in on the relevant portion of the graph. Compare, for example,

```
ezplot(x*exp(-x))
ezplot(x*exp(-x), [-1 4])
```

While you have no direct control over the *y*-domain, `ezplot` attempts to make a reasonable choice.

Entering the command `funtool` (no input arguments) brings up three graphic figures, two of which will display graphs of functions and one containing a control panel. This function calculator lets you manipulate functions and their graphs for pedagogical demonstrations. Type `help funtool` for the details.

## 18.5 Symbolic Matrix Operations

This toolbox lets you represent matrices in symbolic form, as well as MATLAB's numeric form. Given the numeric matrix

```
a  = magic(3)
a  =
     8 1 6
     3 6 7
     4 9 2
```

the function `sym(a)` converts *a* to the symbolic matrix

```
A = sym(a)
A =
[8, 1, 6]
[3, 5, 7]
[4, 9, 2]
```

and `numeric(A)` converts the symbolic matrix back to a numeric one.

Symbolic matrices can also be generated by sym. Try, for example,

```
syms a b t
K = [a+b, a-b; b-a, a+b]
G = [cos(t), sin(t); -sin(t), cos(t)]
```

Here the second matrix $G$ is a Givens rotation matrix.

Algebraic matrix operations with symbolic matrices are computed as you would in MATLAB

| | |
|---|---|
| `K+G` | matrix addition |
| `K-G` | matrix subtraction |
| `K*G` | matrix multiplication |
| `inv(G)` | matrix inversion |
| `K/G or K*inv(G)` | right matrix division |
| `G^2` | power (2) of a matrix |
| `G.'` | transpose of a matrix |
| `G'` | conjugate transpose (Hermitian) of a matrix |

These operations are illustrated by the following, which use the matrices $K$ and $G$ generated above.

```
L = K^2, collect(L), factor(L)
diff(L,a), int(K,a)
J = K/G, simplify(J*G)
simplify(G*(G.'))
```

Note that the initial result of the basic operations may not be in the form desired for your application, so may require further processing with simplify, collect, factor, or expand. These functions, as well as diff and int, act entry-wise on a symbolic matrix.

## 18.6 Symbolic Linear Algebraic Functions

The primary symbolic matrix functions are

| | |
|---|---|
| `det` | determinant |
| `.'` | transpose |
| `'` | Hermitian (conjugate transpose) |
| `inv` | inverse |
| `null` | basis for nullspace |
| `colspace` | basis for column space |
| `eig` | eigenvalues and eigenvectors |
| `poly` | characteristic polynomial |
| `svd` | singular value decomposition |
| `jordan` | Jordan form |

These functions will take either symbolic or numeric arguments, but the output is always symbolic.

Computations with symbolic rational matrices can be carried out exactly. Try, for example,

```
c = floor(10*rand(4)), C = sym(c)
A = inv(C), inv(A), det(A)
b = ones(1,4), x = b/A
x*A
A^3
```

These functions can, of course, be applied to general symbolic matrices. For the matrices $K$ and $G$ defined in the previous section,

```
syms a b t
K = [a+b, a-b; b-a, a+b];
G = [cos(t), sin(t); -sin(t), cos(t)];
```

try, for example (see section 18.7 regarding `solve`),

```
inv(K)
simplify(inv(G))
p = poly(G), simplify(p)
factor(p), X = solve(p)
for j = 1:4, X = simple(X), end
pretty(X)
e = eig(G)
for j = 1:4, e = simple(e); end
pretty(e)
syms t real % make t a real variable
s = svd(G)
for j = 1:4, s = simple(s); end
pretty(s)
```

A typical exercise in a linear algebra course is to determine those values of $t$ so that, say,

$$A = \begin{bmatrix} t & 1 & 0 \\ 1 & t & 1 \\ 0 & 1 & t \end{bmatrix}$$

is singular. The following simple computation shows that this occurs for $t = 0, \sqrt{2}, \sqrt{-2}$.

```
>> syms t
>> A = [t 1 0; 1 t 1; 0 1 t]

A =
[t, 1, 0]
[1, t, 1]
[0, 1, t]
>> p = det(A), solve(p)
p =
```

```
t^3-2*t
ans =
[      0 ]
[2^(1/2)]
[-2^(1/2)]
```

The function eig attempts to compute the eigenvalues and eigenvectors in an exact closed form. Try, for example,

```
A = sym(magic(4)),[V,D] = eig(A)
A = sym(magic(5)),[V,D] = eig(A)
A = sym(magic(6)),[V,D] = eig(A)
```

Except in special cases, however, the result is usually too complicated to be useful. Try, for example, executing

```
A = sym(floor(10*rand(3)))
[V,D] = eig(A)
```

a few times. For this reason, it is usually more efficient to do the computation in variable-precision arithmetic, as is illustrated by

```
A = floor(10*rand(3));
[V,D] = eig(A);
vpa(V), vpa(D)
```

The comments above regarding eig apply as well to the computation of the singular values of a matrix by svd, as can be observed by repeating some of the computations above using svd instead of eig.

## 18.7 Solving Algebraic Equations (solve)

For a symbolic expression $S$, solve(S) will attempt to find the values of the symbolic variable for which the symbolic expression is zero. If an exact symbolic solution is indeed found, you can convert it to a floating point solution, if desired. If an exact symbolic solution cannot be found, then a variable precision one is computed. Moreover, if you have an expression that contains several symbolic variables, you can solve for a particular variable by including it as an input argument in solve.

To solve an equation whose right hand side is not 0, use a quoted string. Try, for example,

```
syms x y z
X = solve(cos(x)+tan(x)); pretty(X)
double(X), vpa(X)
Y = solve(cos(x)-x)
Z = solve(x^2+2*x-1); pretty(Z)
a = solve(x^2+y^2+z^2 + x*y*z);
% a is a solution in the variable x
pretty(a)
b = solve(x^2+y^2+z^2+x*y*z,y);
% b is a solution in y
pretty(b)
```

Symbolic equations are also acceptable to solve. Some examples are

```
X = solve('log(x) = x - 2'), vpa(X)
X = solve('2^x = x + 2'), vpa(X)
% solve for the variable a
A = solve('1 + (a+b)/(a-b) = b','a')
% solve for the variable b
f = solve('1 + (a+b)/(a-b) = b','b')
```

The function solve can also compute the solutions of systems of general algebraic equations. To solve, for example, the nonlinear system below it is convenient to first express the equations as symbolic variables.

```
S1 = 'x^2 + y^2 + z^2 = 2'
S2 = 'x + y = 1'
S3 = 'y + z = 1'
```

The solutions are then computed by

```
[X,Y,Z] = solve(S1, S2, S3)
```

If you alter S2 to

```
S2 = 'x + y + z = 1'
```

then the solution computed by

```
[X,Y,Z] = solve(S1, S2, S3)
```

will be given in terms of square roots.

## 18.8 *Solving Differential Equations* (dsolve)

The function dsolve attempts to solve ordinary differential equations.

The symbolic differential operator is *D*, so that

```
Y = dsolve('Dy = x^2*y', 'x')
```

produces the solution exp(1/3*x^3)*C1 to the differential equation $y' = x^2 y$ . The solution to an initial value problem can be computed by adding a second symbolic expression giving the initial condition.

```
Y = dsolve('Dy = x^2*y','y(0) = 4','x')
```

Notice that in both examples above, the final input argument, `'x'`, is the independent variable of the differential equation. If no independent variable is supplied to `dsolve`, then it is assumed to be *t*. The higher order symbolic differential operators *D2, D3,*.... can be used to solve higher order equations. Explore the following.

```
dsolve('D2y + y = 0')
dsolve('D2y + y = x^2', 'x')
dsolve('D2y+y=x^2','y(0)=4','Dy(0)=1', 'x')
dsolve('D2y – Dy = 2*y')
dsolve('D2y + 6*Dy = 13*y')
Y = dsolve('D2y + 6*Dy + 13*y = cos(t)');
Y = simple(Y)
dsolve('D3y – 3*Dy = 2*y')
pretty(ans)
```

Systems of differential equations can also be solved. For example, to solve the system, it is convenient to first assign the equations to symbolic variables

```
E1 = 'Dx = –2*x + y'
E2 = 'Dy = x – 2*y + z'
E3 = 'Dz = y – 2*z'
```

The solutions are then computed with

```
[x,y,z] = dsolve(E1,E2,E3)
pretty(x), pretty(y), pretty(z)
```

You can explore further details with `help dsolve`.

## 18.9 Further Maple Access

Over fifty special functions of classical applied mathematics are available in this toolbox. Enter `help mfunlist` to see a list of them. These functions can be accessed with the function `mfun`, for which you are referred to `help mfun` for further details.

For many Maple functions, this toolbox provides conversions between MATLAB and Maple syntax to provide access to Maple in a MATLAB environment. To enable you to access much of the remaining functionality of Maple, the function `maple` is provided. It can be used to transmit almost any Maple statement (using Maple syntax) or a Maple function to Maple. You are referred to `help maple` for the details on its use.

The statement `mhelp` *topic* displays Maple's help text for the specified topic.

The Extended Symbolic Math Toolbox provides access to a number of Maple's specialized libraries of procedures. It also provides for use of Maple programming features, although these are usually unnecessary since they duplicate features of MATLAB.

# 19. Subject Area Lists of Functions

There are many MATLAB functions and features that cannot be included in these introductory notes. Listed in the following tables are some of the MATLAB functions and operators, grouped by subject area.[1] You can browse through these lists and use the online help facility, or consult the online *Reference Guide* for more detailed information on the functions.

There are many functions beyond these. In particular, there are specialized collections of M-files for solving particular classes of problems, which are called "Toolboxes"[2]. One of these, the Symbolic Math Toolbox, was introduced in the previous section.

If you enter `ver` alone, you can check the displayed list of topics to determine which toolboxes are available on your local system. These toolboxes can be explored via the command `help` or the local printed documentation.

Toolboxes represent the efforts of some of the world's top researchers, and the list of toolboxes available at the time of printing is:

- Communications Toolbox
- Control System Toolbox
- Financial Toolbox

---

1. Source: MATLAB 5 online *Reference Guide*; *Using MATLAB*; *Symbolic Math Toolbox 2.0 User's Guide*
2. Toolboxes, are optional, so those available to you will depend on your local installation.

- Frequency Domain System Identification Toolbox
- Fuzzy Logic Toolbox
- Higher-Order Spectral Analysis Toolbox
- Image Processing Toolbox
- LMI Control Toolbox
- Mapping Toolbox
- Model Predictive Control Toolbox
- μ-Analysis and Synthesis Toolbox
- NAG® Foundation Toolbox
- Neural Network Toolbox
- Optimization Toolbox
- Partial Differential Equation Toolbox
- QFT Control Design Toolbox
- Robust Control Toolbox
- Signal Processing Toolbox
- Spline Toolbox
- Statistics Toolbox
- Symbolic Math and Extended Symbolic Math Toolboxes
- System Identification Toolbox
- Wavelet Toolbox

Also available is Simulink® an interactive system for modeling and simulating dynamic nonlinear systems in a graphical, mouse-driven environment.

Similar to MATLAB toolboxes, Simulink has domain specific add-ons called "Blocksets". The list of blocksets available at the time of printing is:

- DSP Blockset
- Fixed Point Blockset
- Nonlinear Control Design Blockset

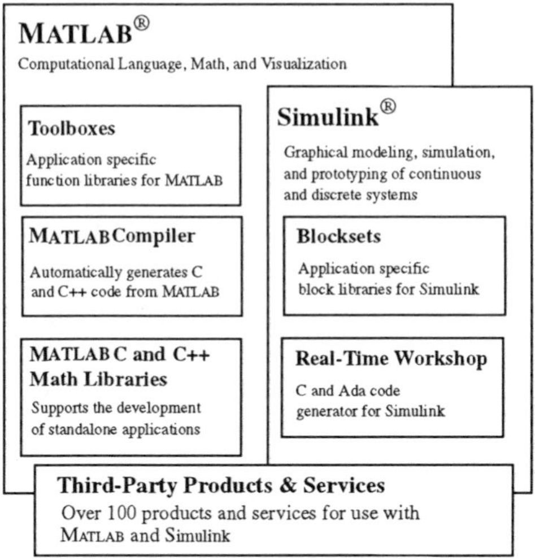

## 19.1 Help Topics - MATLAB Directories

Typing `help` at the MATLAB command prompt will provide a listing of the major MATLAB directories, similar to the following:

| | |
|---|---|
| `matlab\general` | General purpose commands. |
| `matlab\ops` | Operators and special characters. |
| `matlab\lang` | Programming language constructs. |
| `matlab\elmat` | Elementary matrices and manipulation. |
| `matlab\elfun` | Elementary math functions. |
| `matlab\specfun` | Specialized math functions. |
| `matlab\matfun` | Matrix functions - linear algebra. |
| `matlab\datafun` | Data analysis/Fourier transforms. |
| `matlab\polyfun` | Interpolation and polynomials. |
| `matlab\funfun` | Function functions and ODE solvers. |
| `matlab\sparfun` | Sparse matrices. |
| `matlab\graph2d` | Two dimensional graphs. |
| `matlab\graph3d` | Three dimensional graphs. |
| `matlab\specgraph` | Specialized graphs. |
| `matlab\graphics` | Handle Graphics. |
| `matlab\uitools` | Graphical user interface tools. |

| | |
|---|---|
| `matlab\strfun` | Character strings. |
| `matlab\iofun` | File input/output. |
| `matlab\timefun` | Time and dates. |
| `matlab\datatypes` | Data types and structures. |
| `matlab\OS` | OS specific directory. |
| `matlab\demos` | Examples and demonstrations. |
| `toolbox\symbolic` | Symbolic Math Toolbox. |

## 19.2 General Purpose Commands

| Starting and Quitting MATLAB | |
|---|---|
| `matlabrc` | MATLAB startup M-file |
| `quit` | Terminate MATLAB |
| `startup` | MATLAB startup M-file |

| Managing Commands and Functions | |
|---|---|
| `addpath` | Add directories to MATLAB's search path |
| `doc` | Load hypertext documentation |
| `help` | Online help for MATLAB functions and M-files |
| `lasterr` | Last error message |
| `lookfor` | Keyword search through all help entries |
| `path` | Control MATLAB's directory search path |
| `profile` | Measure and display M-file execution profiles |
| `rmpath` | Remove directories from MATLAB's search path |
| `type` | List file |
| `version` | MATLAB version number |
| `what` | Directory listing of M-files, MAT-files, and MEX-files |

| Managing Commands and Functions (Continued) | |
|---|---|
| `whatsnew` | Display README files for MATLAB and toolboxes |
| `which` | Locate functions and files |

| Managing Variables and the Workspace | |
|---|---|
| `clear` | Remove items from memory |
| `disp` | Display text or array |
| `length` | Length of vector |
| `load` | Retrieve variables from disk |
| `pack` | Consolidate workspace memory |
| `save` | Save workspace variables on disk |
| `size` | Array dimensions |
| `who, whos` | List directory of variables in memory |

| Controlling the Command Window | |
|---|---|
| `echo` | Echo M-files during execution |
| `format` | Control the output display format |
| `more` | Control paged output for the command window |

| Working with Files and the Operating Environment | |
|---|---|
| cd | Change working directory |
| delete | Delete files and graphics objects |
| diary | Save session in a disk file |
| dir | Directory listing |
| edit | Edit an M-file |
| fileparts | Filename parts |
| fullfile | Build full filename from parts |
| inmem | Functions in memory |
| matlabroot | Root directory of MATLAB installation |
| tempdir | Return the name of the system's temp directory |
| tempname | Unique name for temporary file |
| ! | Execute operating system command |

## 19.3 Operators and Special Characters

| Operators and Special Characters | |
|---|---|
| + | Plus |
| – | Minus |
| * | Matrix multiplication |
| .* | Array multiplication |
| ^ | Matrix power |

<table>
<tr><th colspan="2">Operators and Special Characters (Continued)</th></tr>
<tr><td>.^</td><td>Array power</td></tr>
<tr><td>kron</td><td>Kronecker tensor product</td></tr>
<tr><td>\</td><td>Backslash or left division</td></tr>
<tr><td>/</td><td>Slash or right division</td></tr>
<tr><td>./ and .\</td><td>Array division, right and left</td></tr>
<tr><td>:</td><td>Colon</td></tr>
<tr><td>( )</td><td>Parentheses</td></tr>
<tr><td>[ ]</td><td>Brackets</td></tr>
<tr><td>{}</td><td>Curly braces</td></tr>
<tr><td>.</td><td>Decimal point</td></tr>
<tr><td>...</td><td>Continuation</td></tr>
<tr><td>,</td><td>Comma</td></tr>
<tr><td>;</td><td>Semicolon</td></tr>
<tr><td>%</td><td>Comment</td></tr>
<tr><td>!</td><td>Exclamation point</td></tr>
<tr><td>'</td><td>Transpose and quote</td></tr>
<tr><td>.'</td><td>Nonconjugated transpose</td></tr>
<tr><td>=</td><td>Assignment</td></tr>
<tr><td>==</td><td>Equality</td></tr>
<tr><td>< ></td><td>Relational operators</td></tr>
<tr><td>&</td><td>Logical AND</td></tr>
<tr><td>|</td><td>Logical OR</td></tr>
</table>

| Operators and Special Characters (Continued) | |
|---|---|
| ~ | Logical NOT |
| xor | Logical exclusive (XOR) |

| Logical Functions | |
|---|---|
| all | Test to determine if all elements are nonzero |
| any | Test for any nonzeros |
| exist | Check if a variable or file exists |
| find | Find indices and values of nonzero elements |
| is* | Detect state |
| isa | Detect an object of a given class |
| logical | Convert numeric values to logical |

| Bitwise Functions | |
|---|---|
| bitand | Bit-wise AND |
| bitcmp | Complement bits |
| bitor | Bit-wise OR |
| bitmax | Maximum floating-point integer |
| bitset | Set bit |
| bitshift | Bit-wise shift |
| bitget | Get bit |
| bitxor | Bit-wise XOR |

## 19.4 Language Constructs and Debugging

| MATLAB as a Programming Language | |
|---|---|
| `builtin` | Execute builtin function from overloaded method |
| `eval` | Interpret strings containing MATLAB expressions |
| `feval` | Function evaluation |
| `function` | Function M-files |
| `global` | Define global variables |
| `nargchk` | Check number of input arguments |
| `script` | Script M-files |

| Control Flow | |
|---|---|
| `break` | Break out of flow control structures |
| `case` | Case switch |
| `else` | Conditionally execute statements |
| `elseif` | Conditionally execute statements |
| `end` | Terminate for, while, switch, and if statements or indicate last index |
| `error` | Display error messages |
| `for` | Repeat statements a specific number of times |
| `if` | Conditionally execute statements |
| `otherwise` | Default part of switch statement |

| **Control Flow (Continued)** | |
|---|---|
| `return` | Return to the invoking function |
| `switch` | Switch among several cases based on expression |
| `warning` | Display warning message |
| `while` | Repeat statements an indefinite number of times |

| **Interactive Input** | |
|---|---|
| `input` | Request user input |
| `keyboard` | Invoke the keyboard in an M-file |
| `menu` | Generate a menu of choices for user input |
| `pause` | Halt execution temporarily |

| **Debugging** | |
|---|---|
| `dbclear` | Clear breakpoints |
| `dbcont` | Resume execution |
| `dbdown` | Change local workspace context |
| `dbmex` | Enable MEX-file debugging |
| `dbquit` | Quit debug mode |
| `dbstack` | Display function call stack |
| `dbstatus` | List all breakpoints |

| Debugging (Continued) | |
|---|---|
| dbstep | Execute one or more lines from a breakpoint |
| dbstop | Set breakpoints in an M-file function |
| dbtype | List M-file with line numbers |
| dbup | Change local workspace context |

## 19.5 Elementary Matrices and Matrix Manipulation

| Elementary Matrices and Arrays | |
|---|---|
| eye | Identity matrix |
| linspace | Generate linearly spaced vectors |
| logspace | Generate logarithmically spaced vectors |
| ones | Create an array of all ones |
| rand | Uniformly distributed random numbers and arrays |
| randn | Normally distributed random numbers and arrays |
| zeros | Create an array of all zeros |
| : (colon) | Regularly spaced vector |

| Special Variables and Constants | |
|---|---|
| `ans` | The most recent answer |
| `computer` | Identify the computer on which MATLAB is running |
| `eps` | Floating-point relative accuracy |
| `flops` | Count floating-point operations |
| `i` | Imaginary unit |
| `Inf` | Infinity |
| `inputname` | Input argument name |
| `j` | Imaginary unit |
| `NaN` | Not-a-Number |
| `nargin`, `nargout` | Number of function arguments |
| `pi` | Ratio of a circle's circumference to its diameter |
| `realmax` | Largest positive floating-point number |
| `realmin` | Smallest positive floating-point number |
| `varargin`, `varargout` | Pass or return variable numbers of arguments |

| Time and Dates | |
|---|---|
| `calendar` | Calendar |
| `clock` | Current time as a date vector |
| `cputime` | Elapsed CPU time |

| Time and Dates (Continued) | |
|---|---|
| date | Current date string |
| datenum | Serial date number |
| datestr | Date string format |
| datevec | Date components |
| eomday | End of month |
| etime | Elapsed time |
| now | Current date and time |
| tic, toc | Stopwatch timer |
| weekday | Day of the week |

| Matrix Manipulation | |
|---|---|
| cat | Concatenate arrays |
| diag | Diagonal matrices and diagonals of a matrix |
| fliplr | Flip matrices left-right |
| flipud | Flip matrices up-down |
| repmat | Replicate and tile an array |
| reshape | Reshape array |
| rot90 | Rotate matrix 90 degrees |
| tril | Lower triangular part of a matrix |
| triu | Upper triangular part of a matrix |
| : (colon) | Index into array, rearrange array |

| Specialized Matrices | |
|---|---|
| compan | Companion matrix |
| gallery | Test matrices |
| hadamard | Hadamard matrix |
| hankel | Hankel matrix |
| hilb | Hilbert matrix |
| invhilb | Inverse of the Hilbert matrix |
| magic | Magic square |
| pascal | Pascal matrix |
| toeplitz | Toeplitz matrix |
| wilkinson | Wilkinson's eigenvalue test matrix |

## 19.6 Elementary Math Functions

| Elementary Math Functions | |
|---|---|
| abs | Absolute value and complex magnitude |
| acos, acosh | Inverse cosine and inverse hyperbolic cosine |
| acot, acoth | Inverse cotangent and inverse hyperbolic cotangent |
| acsc, acsch | Inverse cosecant and inverse hyperbolic cosecant |
| angle | Phase angle |

| Elementary Math Functions (Continued) | |
|---|---|
| asec, asech | Inverse secant and inverse hyperbolic secant |
| asin, asinh | Inverse sine and inverse hyperbolic sine |
| atan, atanh | Inverse tangent and inverse hyperbolic tangent |
| atan2 | Four quadrant inverse tangent |
| ceil | Round toward infinity |
| conj | Complex conjugate |
| cos, cosh | Cosine and hyperbolic cosine |
| cot, coth | Cotangent and hyperbolic cotangent |
| csc, csch | Cosecant and hyperbolic cosecant |
| exp | Exponential |
| fix | Round towards zero |
| floor | Round towards minus infinity |
| gcd | Greatest common divisor |
| imag | Imaginary part of a complex number |
| lcm | Least common multiple |
| log | Natural logarithm |
| log2 | Base 2 logarithm, dissect floating-point numbers into exponent and mantissa |
| log10 | Common (base 10) logarithm |
| mod | Modulus (signed remainder after division) |
| real | Real part of complex number |
| rem | Remainder after division |

| Elementary Math Functions (Continued) | |
|---|---|
| `round` | Round to nearest integer |
| `sec, sech` | Secant and hyperbolic secant |
| `sign` | Signum function |
| `sin, sinh` | Sine and hyperbolic sine |
| `sqrt` | Square root |
| `tan, tanh` | Tangent and hyperbolic tangent |

## 19.7 Specialized Math Functions

| Specialized Math Functions | |
|---|---|
| `airy` | Airy functions |
| `besselh` | Bessel functions of the third kind (Hankel functions) |
| `besseli, besselk` | Modified Bessel functions |
| `besselj, bessely` | Bessel functions |
| `beta, betainc, betaln` | Beta functions |
| `ellipj` | Jacobi elliptic functions |
| `ellipke` | Complete elliptic integrals of the first and second kind |
| `erf, erfc, erfcx, erfinv` | Error functions |

| Specialized Math Functions (Continued) | |
|---|---|
| `expint` | Exponential integral |
| `gamma`, `gammainc`, `gammaln` | Gamma functions |
| `legendre` | Associated Legendre functions |
| `pow2` | Base 2 power, scale floating-point numbers |
| `rat`, `rats` | Rational fraction approximation |

| Coordinate System Conversion | |
|---|---|
| `cart2pol` | Transform Cartesian coordinates to polar or cylindrical |
| `cart2sph` | Transform Cartesian coordinates to spherical |
| `pol2cart` | Transform polar or cylindrical coordinates to Cartesian |
| `sph2cart` | Transform spherical coordinates to Cartesian |

## 19.8 Matrix Functions – Numerical Linear Algebra

| Matrix Analysis | |
|---|---|
| `cond` | Condition number with respect to inversion |
| `condeig` | Condition number with respect to eigenvalues |
| `det` | Matrix determinant |
| `norm` | Vector and matrix norms |
| `null` | Null space of a matrix |
| `orth` | Range space of a matrix |
| `rank` | Rank of a matrix |
| `rcond` | Matrix reciprocal condition number estimate |
| `rref`, `rrefmovie` | Reduced row echelon form |
| `subspace` | Angle between two subspaces |
| `trace` | Sum of diagonal elements |

| Linear Equations | |
|---|---|
| \, / | Linear equation solution |
| `chol` | Cholesky factorization |
| `inv` | Matrix inverse |
| `lscov` | Least squares solution in the presence of known covariance |
| `lu` | LU matrix factorization |

| Linear Equations (Continued) | |
|---|---|
| `nnls` | Nonnegative least squares |
| `pinv` | Moore-Penrose pseudoinverse of a matrix |
| `qr` | Orthogonal-triangular decomposition |

| Eigenvalues and Singular Values | |
|---|---|
| `balance` | Improve accuracy of computed eigenvalues |
| `cdf2rdf` | Convert complex diagonal form to real block diagonal form |
| `eig` | Eigenvalues and eigenvectors |
| `hess` | Hessenberg form of a matrix |
| `poly` | Polynomial with specified roots |
| `qz` | QZ factorization for generalized eigenvalues |
| `rsf2csf` | Convert real Schur form to complex Schur form |
| `schur` | Schur decomposition |
| `svd` | Singular value decomposition |

| Matrix Functions | |
|---|---|
| `expm` | Matrix exponential |
| `funm` | Evaluate functions of a matrix |

| Matrix Functions (Continued) | |
|---|---|
| logm | Matrix logarithm |
| sqrtm | Matrix square root |

| Low Level Functions | |
|---|---|
| qrdelete | Delete column from QR factorization |
| qrinsert | Insert column in QR factorization |

## 19.9 Data Analysis and Fourier Transform Functions

| Basic Operations | |
|---|---|
| convhull | Convex hull |
| cumprod | Cumulative product |
| cumsum | Cumulative sum |
| cumtrapz | Cumulative trapezoidal numerical integration |
| delaunay | Delaunay triangulation |
| dsearch | Search for nearest point |
| factor | Prime factors |
| inpolygon | Detect points inside a polygonal region |
| max | Maximum elements of an array |
| mean | Average or mean value of arrays |
| median | Median value of arrays |

| Basic Operations (Continued) | |
|---|---|
| min | Minimum elements of an array |
| perms | All possible permutations |
| polyarea | Area of polygon |
| primes | Generate list of prime numbers |
| prod | Product of array elements |
| sort | Sort elements in ascending order |
| sortrows | Sort rows in ascending order |
| std | Standard deviation |
| sum | Sum of array elements |
| trapz | Trapezoidal numerical integration |
| tsearch | Search for enclosing Delaunay triangle |
| voronoi | Voronoi diagram |

| Finite Differences | |
|---|---|
| del2 | Discrete Laplacian |
| diff | Differences and approximate derivatives |
| gradient | Numerical gradient |

| Correlation | |
|---|---|
| corrcoef | Correlation coefficients |
| cov | Covariance matrix |

| Filtering and Convolution | |
|---|---|
| conv | Convolution and polynomial multiplication |
| conv2 | Two-dimensional convolution |
| deconv | Deconvolution and polynomial division |
| filter | Filter data with an infinite impulse response (IIR) or finite impulse response (FIR) filter |
| filter2 | Two-dimensional digital filtering |

| Fourier Transforms | |
|---|---|
| abs | Absolute value and complex magnitude |
| angle | Phase angle |
| cplxpair | Sort complex numbers into complex conjugate pairs |
| fft | One-dimensional fast Fourier transform |
| fft2 | Two-dimensional fast Fourier transform |
| fftshift | Shift DC component of fast Fourier transform to center of spectrum |
| ifft | Inverse one-dimensional fast Fourier transform |
| ifft2 | Inverse two-dimensional fast Fourier transform |
| nextpow2 | Next power of two |
| unwrap | Correct phase angles |

| Vector Functions | |
|---|---|
| cross | Vector cross product |
| intersect | Set intersection of two vectors |
| ismember | Detect members of a set |
| setdiff | Return the set difference of two vectors |
| setxor | Set exclusive-or of two vectors |
| union | Set union of two vectors |
| unique | Unique elements of a vector |

## 19.10 Polynomial and Interpolation Functions

| Polynomials | |
|---|---|
| conv | Convolution and polynomial multiplication |
| deconv | Deconvolution and polynomial division |
| poly | Polynomial with specified roots |
| polyder | Polynomial derivative |
| polyeig | Polynomial eigenvalue problem |
| polyfit | Polynomial curve fitting |
| polyval | Polynomial evaluation |
| polyvalm | Matrix polynomial evaluation |
| residue | Convert between partial fraction expansion and polynomial coefficients |
| roots | Polynomial roots |

| **Data Interpolation** | |
|---|---|
| `griddata` | Data gridding |
| `interp1` | One-dimensional data interpolation (table lookup) |
| `interp2` | Two-dimensional data interpolation (table lookup) |
| `interp3` | Three-dimensional data interpolation (table lookup) |
| `interpft` | One-dimensional interpolation using the FFT method |
| `interpn` | Multidimensional data interpolation (table lookup) |
| `meshgrid` | Generate X and Y matrices for three-dimensional plots |
| `ndgrid` | Generate arrays for multidimensional functions and interpolation |
| `spline` | Cubic spline interpolation |

## 19.11 Function Functions – Nonlinear Numerical Methods

| Function Functions – Nonlinear Numerical Methods | |
|---|---|
| dblquad | Numerical double integration |
| fmin | Minimize a function of one variable |
| fmins | Minimize a function of several variables |
| fzero | Zero of a function of one variable |
| ode45, ode23, ode113, ode15s, ode23s | Solve differential equations |
| odefile | Define a differential equation problem for ODE solvers |
| odeget | Extract properties from options structure created with odeset |
| odeset | Create or alter options structure for input to ODE solvers |
| quad, quad8 | Numerical evaluation of integrals |
| vectorize | Vectorize expression |

## 19.12 Sparse Matrix Functions

| Elementary Sparse Matrices | |
|---|---|
| `spdiags` | Extract and create sparse band and diagonal matrices |
| `speye` | Sparse identity matrix |
| `sprand` | Sparse uniformly distributed random matrix |
| `sprandn` | Sparse normally distributed random matrix |
| `sprandsym` | Sparse symmetric random matrix |

| Full to Sparse Conversion | |
|---|---|
| `find` | Find indices and values of nonzero elements |
| `full` | Convert sparse matrix to full matrix |
| `sparse` | Create sparse matrix |
| `spconvert` | Import matrix from sparse matrix external format |

| Working with Nonzero Entries of Sparse Matrices | |
|---|---|
| `nnz` | Number of nonzero matrix elements |
| `nonzeros` | Nonzero matrix elements |
| `nzmax` | Amount of storage allocated for nonzero matrix elements |

| **Working with Nonzero Entries of Sparse Matrices (Continued)** | |
|---|---|
| `spalloc` | Allocate space for sparse matrix |
| `spfun` | Apply function to nonzero sparse matrix elements |
| `spones` | Replace nonzero sparse matrix elements with ones |

| **Reordering Algorithms** | |
|---|---|
| `colmmd` | Sparse column minimum degree permutation |
| `colperm` | Sparse column permutation based on nonzero count |
| `dmperm` | Dulmage-Mendelsohn decomposition |
| `randperm` | Random permutation |
| `symmmd` | Sparse symmetric minimum degree ordering |
| `symrcm` | Sparse reverse Cuthill-McKee ordering |

| **Norm, Condition Number, and Rank** | |
|---|---|
| `condest` | 1–norm matrix condition number estimate |
| `normest` | 2–norm estimate |

| **Sparse Systems of Linear Equations** | |
|---|---|
| bicg | BiConjugate Gradients method |
| bicgstab | BiConjugate Gradients Stabilized method |
| cgs | Conjugate Gradients Squared method |
| cholinc | Incomplete Cholesky factorizations |
| gmres | Generalized Minimum Residual method (with restarts) |
| luinc | Incomplete LU matrix factorizations |
| pcg | Preconditioned Conjugate Gradients method |
| qmr | Quasi-Minimal Residual method |

| **Miscellaneous** | |
|---|---|
| spparms | Set parameters for sparse matrix routines |
| spy | Visualize sparsity pattern |

| **Sparse Eigenvalues and Singular Values** | |
|---|---|
| eigs | Find a few eigenvalues and eigenvectors |
| svds | A few singular values |

## 19.13 Sound Processing Functions

| General and Platform Specific Sound Functions | |
|---|---|
| sound | Convert vector into sound |
| auread | Read NeXT/SUN (.au) sound file |
| auwrite | Write NeXT/SUN (.au) sound file |
| wavread | Read Microsoft WAVE (.wav) sound file |
| wavwrite | Write Microsoft WAVE (.wav) sound file |
| readsnd | Read snd resources and files |
| recordsound | Record sound |
| soundcap | Sound capabilities |
| speak | Speak text string |
| writesnd | Write snd resources and files |

## 19.14 Graph2d – Two Dimensional Plotting

| Elementary X-Y graphs | |
|---|---|
| plot | Linear plot |
| loglog | Log-log scale plot |
| semilogx | Semi-log scale plot |
| semilogy | Semi-log scale plot |
| polar | Polar coordinate plot |
| plotyy | Graphs with y tick labels on the left and right |

| **Axis control** | |
|---|---|
| `axis` | Control axis scaling and appearance |
| `zoom` | Zoom in and out on a 2–D plot |
| `grid` | Grid lines |
| `box` | Axis box |
| `hold` | Hold current graph |
| `axes` | Create axes in arbitrary positions |
| `subplot` | Create axes in tiled positions |

| **Graph annotation** | |
|---|---|
| `legend` | Graph legend |
| `title` | Graph title |
| `xlabel` | X–axis label |
| `ylabel` | Y–axis label |
| `text` | Text annotation |
| `gtext` | Place text with mouse |

| **Hardcopy and printing** | |
|---|---|
| `print` | Print graph or SIMULINK system; or save graph to M-file |
| `printopt` | Printer defaults |
| `orient` | Set paper orientation |

## 19.15 Graph3d – Three Dimensional Plotting

| Elementary 3-D plots | |
|---|---|
| plot3 | Plot lines and points in 3-D space |
| mesh | 3-D mesh surface |
| surf | 3-D colored surface |
| fill3 | Filled 3-D polygons |

| Color control | |
|---|---|
| colormap | Color look-up table |
| caxis | Pseudocolor axis scaling |
| shading | Color shading mode |
| hidden | Mesh hidden line removal mode |
| brighten | Brighten or darken color map |

| Lighting | |
|---|---|
| surfl | 3-D shaded surface with lighting |
| lighting | Lighting mode |
| material | Material reflectance mode |
| specular | Specular reflectance |
| diffuse | Diffuse reflectance |
| surfnorm | Surface normals |

| **Colormaps** | |
|---|---|
| hsv | Hue-saturation-value color map |
| hot | Black-red-yellow-white color map |
| gray | Linear gray-scale color map |
| bone | Gray-scale with tinge of blue color map |
| copper | Linear copper-tone color map |
| pink | Pastel shades of pink color map |
| white | All white color map |
| flag | Alternating red, white, blue, and black color map |
| lines | Color map with the line colors |
| colorcube | Enhanced color-cube color map |
| jet | Variant of HSV |
| prism | Prism color map |
| cool | Shades of cyan and magenta color map |
| autumn | Shades of red and yellow color map |
| spring | Shades of magenta and yellow color map |
| winter | Shades of blue and green color map |
| summer | Shades of green and yellow color map |

| Axis control | |
|---|---|
| `axis` | Control axis scaling and appearance |
| `zoom` | Zoom in and out on a 2-D plot |
| `grid` | Grid lines |
| `box` | Axis box |
| `hold` | Hold current graph |
| `axes` | Create axes in arbitrary positions |
| `subplot` | Create axes in tiled positions |

| Viewpoint control | |
|---|---|
| `view` | 3-D graph viewpoint specification |
| `viewmtx` | View transformation matrix |
| `rotate3d` | Interactively rotate view of 3-D plot |

| Graph annotation | |
|---|---|
| `title` | Graph title |
| `xlabel` | X-axis label |
| `ylabel` | Y-axis label |
| `zlabel` | Z-axis label |
| `colorbar` | Display color bar (color scale) |
| `text` | Text annotation |
| `gtext` | Mouse placement of text |

## 19.16 Specgraph – Specialized Graphs

| Specialized 2-D graphs | |
|---|---|
| area | Filled area plot |
| bar | Bar graph |
| barh | Horizontal bar graph |
| bar3 | 3-D bar graph |
| bar3h | Horizontal 3-D bar graph |
| comet | Comet-like trajectory |
| errorbar | Error bar plot |
| ezplot | Easy to use function plotter |
| feather | Feather plot |
| fill | Filled 2-D polygons |
| fplot | Plot function |
| hist | Histogram |
| pareto | Pareto chart |
| pie | Pie chart |
| pie3 | 3-D pie chart |
| plotmatrix | Scatter plot matrix |
| ribbon | Draw 2-D lines as ribbons in 3-D |
| stem | Discrete sequence or "stem" plot |
| stairs | Stairstep plot |
| area | Filled area plot |

| Contour and 2–1/2 D graphs | |
|---|---|
| `contour` | Contour plot |
| `contourf` | Filled contour plot |
| `contour3` | 3-D Contour plot |
| `clabel` | Contour plot elevation labels |
| `pcolor` | Pseudocolor (checkerboard) plot |
| `quiver` | Quiver plot |
| `voronoi` | Voronoi diagram |

| Specialized 3-D graphs | |
|---|---|
| `comet3` | 3-D comet-like trajectories |
| `meshc` | Combination mesh/contour plot |
| `meshz` | 3-D mesh with curtain |
| `stem3` | 3-D stem plot |
| `quiver3` | 3-D quiver plot |
| `slice` | Volumetric slice plot |
| `surfc` | Combination surf/contour plot |
| `trisurf` | Triangular surface plot |
| `trimesh` | Triangular mesh plot |
| `waterfall` | Waterfall plot |

| Images display and file I/O | |
|---|---|
| `image` | Display image |
| `imagesc` | Scale data and display as image |
| `colormap` | Color look-up table |
| `gray` | Linear gray-scale color map |
| `contrast` | Gray scale color map to enhance image contrast |
| `brighten` | Brighten or darken color map |
| `colorbar` | Display color bar (color scale) |
| `imread` | Read image from graphics file |
| `imwrite` | Write image to graphics file |
| `imfinfo` | Information about graphics file |

| Movies and animation | |
|---|---|
| `capture` | Screen capture of current figure |
| `moviein` | Initialize movie frame memory |
| `getframe` | Get movie frame |
| `movie` | Play recorded movie frames |
| `qtwrite` | Translate movie into QuickTime format (Macintosh only) |
| `rotate` | Rotate object about specified orgin and direction |
| `frame2im` | Convert movie frame to indexed image |
| `im2frame` | Convert index image into movie format |

| **Color related functions** | |
|---|---|
| `spinmap` | Spin color map |
| `rgbplot` | Plot color map |
| `colstyle` | Parse color and style from string |

| **Solid modeling** | |
|---|---|
| `cylinder` | Generate cylinder |
| `sphere` | Generate sphere |
| `patch` | Create patch |

## 19.17 Handle Graphics

| **Figure window creation and control** | |
|---|---|
| `figure` | Create figure window |
| `gcf` | Get handle to current figure |
| `clf` | Clear current figure |
| `shg` | Show graph window |
| `close` | Close figure |
| `refresh` | Refresh figure |

| **Axis creation and control** | |
|---|---|
| `subplot` | Create axes in tiled positions |
| `axes` | Create axes in arbitrary positions |
| `gca` | Get handle to current axes |
| `cla` | Clear current axes |
| `axis` | Control axis scaling and appearance |
| `box` | Axis box |
| `caxis` | Control pseudocolor axis scaling |
| `hold` | Hold current graph |
| `ishold` | Return hold state |

| **Handle Graphics objects** | |
|---|---|
| `figure` | Create figure window |
| `axes` | Create axes |
| `line` | Create line |
| `text` | Create text |
| `patch` | Create patch |
| `surface` | Create surface |
| `image` | Create image |
| `light` | Create light |
| `uicontrol` | Create user interface control |
| `uimenu` | Create user interface menu |

| Handle Graphics operations | |
|---|---|
| `set` | Set object properties |
| `get` | Get object properties |
| `reset` | Reset object properties |
| `delete` | Delete object |
| `gco` | Get handle to current object |
| `gcbo` | Get handle to current callback object |
| `gcbf` | Get handle to current callback figure |
| `drawnow` | Flush pending graphics events |
| `findobj` | Find objects with specified property values |
| `copyobj` | Make copy of graphics object and its children |

| Utilities | |
|---|---|
| `closereq` | Figure close request function |
| `newplot` | M-file preamble for NextPlot property |
| `ishandle` | True for graphics handles |

## 19.18 Uitools – Graphical User Interface Tools

| GUI functions | |
|---|---|
| `uicontrol` | Create user interface control |
| `uimenu` | Create user interface menu |
| `ginput` | Graphical input from mouse |
| `dragrect` | Drag XOR rectangles with mouse |
| `rbbox` | Rubberband box |
| `selectmoveresize` | Interactively select, move, resize, or copy objects |
| `waitforbuttonpress` | Wait for key/buttonpress over figure |
| `waitfor` | Block execution and wait for event |
| `uiwait` | Block execution and wait for resume |
| `uiresume` | Resume execution of blocked M-file |

| GUI design tools | |
|---|---|
| `guide` | Design GUI |
| `align` | Align uicontrols and axes |
| `cbedit` | Edit callback |
| `menuedit` | Edit menu |
| `propedit` | Edit property |

| Dialog boxes | |
|---|---|
| `dialog` | Create dialog figure |
| `axlimdlg` | Axes limits dialog box |
| `errordlg` | Error dialog box |
| `helpdlg` | Help dialog box |
| `inputdlg` | Input dialog box |
| `listdlg` | List selection dialog box |
| `menu` | Generate menu of choices for user input |
| `msgbox` | Message box |
| `questdlg` | Question dialog box |
| `warndlg` | Warning dialog box |
| `uigetfile` | Standard open file dialog box |
| `uiputfile` | Standard save file dialog box |
| `uisetcolor` | Color selection dialog box |
| `uisetfont` | Font selection dialog box |
| `pagedlg` | Page position dialog box |
| `printdlg` | Print dialog box |
| `waitbar` | Display wait bar |

| Menu utilities | |
|---|---|
| `makemenu` | Create menu structure |
| `menubar` | Computer dependent default setting for MenuBar property |

| Menu utilities (Continued) | |
|---|---|
| `umtoggle` | Toggle "checked" status of uimenu object |
| `winmenu` | Create submenu for "Window" menu item |

| Toolbar button group utilities | |
|---|---|
| `btngroup` | Create toolbar button group |
| `btnstate` | Query state of toolbar button group |
| `btnpress` | Button press manager for toolbar button group |
| `btndown` | Depress button in toolbar button group |
| `btnup` | Raise button in toolbar button group |

| User-defined figure/axes property utilities | |
|---|---|
| `clruprop` | Clear user-defined property |
| `getuprop` | Get value of user-defined property |
| `setuprop` | Set user-defined property |

| Miscellaneous utilities | |
|---|---|
| `allchild` | Get all object children |
| `findall` | Find all objects |
| `hidegui` | Hide/unhide GUI |
| `edtext` | Interactive editing of axes text objects |

| Miscellaneous utilities (Continued) | |
|---|---|
| `getstatus` | Get status text string in figure |
| `setstatus` | Set status text string in figure |
| `popupstr` | Get popup menu selection string |
| `remapfig` | Transform figure objects' positions |
| `setptr` | Set figure pointer |
| `getptr` | Get figure pointer |
| `overobj` | Get handle of object the pointer is over |

## 19.19 Character String Functions

| General | |
|---|---|
| `abs` | Absolute value and complex magnitude |
| `eval` | Interpret strings containing MATLAB expressions |
| `real` | Real part of complex number |
| `strings` | MATLAB string handling |

| String Manipulation | |
|---|---|
| `deblank` | Strip trailing blanks from the end of a string |
| `findstr` | Find one string within another |
| `lower` | Convert string to lower case |

| String Manipulation (Continued) | |
|---|---|
| strcat | String concatenation |
| strcmp | Compare strings |
| strjust | Justify a character array |
| strmatch | Find possible matches for a string |
| strncmp | Compare the first n characters of two strings |
| strrep | String search and replace |
| strtok | First token in string |
| strvcat | Vertical concatenation of strings |
| upper | Convert string to upper case |

| String to Number Conversion | |
|---|---|
| char | Create character array (string) |
| int2str | Integer to string conversion |
| mat2str | Convert a matrix into a string |
| num2str | Number to string conversion |
| sprintf | Write formatted data to a string |
| sscanf | Read string under format control |
| str2num | String to number conversion |

| Radix Conversion | |
|---|---|
| bin2dec | Binary to decimal number conversion |
| dec2bin | Decimal to binary number conversion |
| dec2hex | Decimal to hexadecimal number conversion |
| hex2dec | IEEE hexadecimal to decimal number conversion |
| hex2num | Hexadecimal to double number conversion |

## 19.20 Low-Level File I/O Functions

| File Opening and Closing | |
|---|---|
| fclose | Close one or more open files |
| fopen | Open a file or obtain information about open files |

| Unformatted I/O | |
|---|---|
| fread | Read binary data from file |
| fwrite | Write binary data to a file |

| Formatted I/O | |
|---|---|
| fgetl | Return the next line of a file as a string without line terminator(s) |
| fgets | Return the next line of a file as a string with line terminator(s) |
| fprintf | Write formatted data to file |
| fscanf | Read formatted data from file |

| File Positioning | |
|---|---|
| feof | Test for end-of-file |
| ferror | Query MATLAB about errors in file input or output |
| frewind | Rewind an open file |
| fseek | Set file position indicator |
| ftell | Get file position indicator |

| String Conversion | |
|---|---|
| sprintf | Write formatted data to a string |
| sscanf | Read string under format control |

| Specialized File I/O | |
|---|---|
| qtwrite | Write QuickTime movie file to disk |
| dlmread | Read an ASCII delimited file into a matrix |
| qtwrite | Write QuickTime movie file to disk |
| dlmread | Read an ASCII delimited file into a matrix |
| dlmwrite | Write a matrix to an ASCII delimited file |
| imfinfo | Return information about a graphics file |
| imread | Read an image from a graphics file |
| imwrite | Write an image to a graphics file |
| wk1read | Read a Lotus WK1 spreadsheet file into a matrix |
| wk1write | Write a matrix to a Lotus123 WK1 spreadsheet file |
| xlgetrange | Get range of cells from Microsoft Excel worksheet |
| xlsetrange | Set range of cells in Microsoft Excel worksheet |

## 19.21 Data Types and Structures

| Data types (classes) | |
|---|---|
| `double` | Convert to double precision |
| `sparse` | Create sparse matrix |
| `char` | Create character array (string) |
| `cell` | Create cell array |
| `struct` | Create or convert to structure array |
| `uint8` | Convert to unsigned 8-bit integer |

| Multidimensional Array Functions | |
|---|---|
| `cat` | Concatenate arrays |
| `flipdim` | Flip array along a specified dimension |
| `ind2sub` | Subscripts from linear index |
| `ipermute` | Inverse permute the dimensions of a multidimensional array |
| `ndgrid` | Generate arrays for multidimensional functions and interpolation |
| `ndims` | Number of array dimensions |
| `permute` | Rearrange the dimensions of a multidimensional array |
| `reshape` | Reshape array |
| `shiftdim` | Shift dimensions |
| `squeeze` | Remove singleton dimensions |
| `sub2ind` | Single index from subscripts |

| Cell Array Functions | |
|---|---|
| cell | Create cell array |
| cellstr | Create cell array of strings from character array |
| cell2struct | Cell array to structure array conversion |
| celldisp | Display cell array contents. |
| num2cell | Convert a numeric array into a cell array |
| cellplot | Graphically display the structure of cell arrays |

| Structure Functions | |
|---|---|
| fieldnames | Field names of a structure |
| getfield | Get field of structure array |
| rmfield | Remove structure fields |
| setfield | Set field of structure array |
| struct | Create structure array |
| struct2cell | Structure to cell array conversion |

| Object-Oriented Programming | |
|---|---|
| class | Create object or return class of object |
| inferiorto | Inferior class relationship |
| inline | Construct an inline object |

| Object-Oriented Programming | |
|---|---|
| `isa` | Detect an object of a given class |
| `superiorto` | Superior class relationship |

## 19.22 The Symbolic Math Toolbox

| Calculus | |
|---|---|
| `diff` | Differentiate |
| `int` | Integrate |
| `jacobian` | Jacobian matrix |
| `limit` | Limit of an expression |
| `symsum` | Summation of series |
| `taylor` | Taylor series expansion |

| Linear Algebra | |
|---|---|
| `colspace` | Basis for column space |
| `det` | Determinant |
| `diag` | Create or extract diagonals |
| `eig` | Eigenvalues and eigenvectors |
| `expm` | Matrix exponential |
| `inv` | Matrix inverse |
| `jordan` | Jordan canonical form |
| `null` | Basis for null space |

| Linear Algebra (Continued) | |
|---|---|
| poly | Characteristic polynomial |
| rank | Matrix rank |
| rref | Reduced row echelon form |
| svd | Singular value decomposition |
| tril | Lower triangle |
| triu | Upper triangle |

| Simplification | |
|---|---|
| collect | Collect common terms |
| expand | Expand polynomials and elementary functions |
| factor | Factor |
| horner | Nested polynomial representation |
| numden | Numerator and denominator |
| simple | Search for shortest form |
| simplify | Simplification |
| subexpr | Rewrite in terms of subexpressions |

| **Solution of Equations** | |
|---|---|
| compose | Functional composition |
| dsolve | Solution of differential equations |
| finverse | Functional inverse |
| solve | Solution of algebraic equations |

| **Variable Precision Arithmetic** | |
|---|---|
| digits | Set variable precision accuracy |
| vpa | Variable precision arithmetic |

| **Arithmetic Operations** | |
|---|---|
| + | Addition |
| - | Subtraction |
| * | Multiplication |
| .* | Array multiplication |
| / | Right division |
| ./ | Array right division |
| \ | Left division |
| .\ | Array left division |
| ^ | Matrix or scalar raised to a power |
| .^ | Array raised to a power |

| Arithmetic Operations (Continued) | |
|---|---|
| ' | Complex conjugate transpose |
| .' | Real transpose |

| Special Functions | |
|---|---|
| cosint | Cosine integral, $Ci(x)$ |
| lambertw | Solution of $\lambda(x)e^{\lambda(x)} = x$ |
| sinint | Sine integral, $Si(x)$ |
| zeta | Riemann zeta function |

| Access To Maple | |
|---|---|
| maple | Access Maple kernel |
| mapleinit | Initialize Maple |
| mfun | Numeric evaluation of Maple functions |
| mhelp | Maple help |
| mfunlist | List of functions for mfun |
| procread | Install a Maple procedure |

| Pedagogical and Graphical Applications | |
|---|---|
| ezplot | Easy to use function plotter |
| funtool | Function calculator |
| rsums | Riemann sums |

| **Conversions** | |
|---|---|
| `char` | Convert sym object to string |
| `double` | Convert symbolic matrix to double |
| `poly2sym` | Function calculator |
| `sym2poly` | Symbolic polynomial to coefficient vector |

| **Basic Operations** | |
|---|---|
| `ccode` | C code representation of a symbolic expression |
| `conj` | Complex conjugate |
| `findsym` | Determine symbolic variables |
| `fortran` | Fortran representation of a symbolic expression |
| `imag` | Imaginary part of a complex number |
| `latex` | LaTeX representation of a symbolic expression |
| `pretty` | Pretty print a symbolic expression |
| `real` | Real part of an imaginary number |
| `sym` | Create symbolic object |
| `syms` | Shortcut for creating multiple symbolic objects |

| Integral Transforms | |
|---|---|
| `fourier` | Fourier integral transform |
| `ifourier` | Inverse Fourier integral transform |
| `ilaplace` | Inverse Laplace transform |
| `iztrans` | Inverse $z$-transform |
| `laplace` | Laplace transform |
| `ztrans` | $z$-transform |

# Index

## D



## E



## F